THE SMART GUIDE TO

Biology

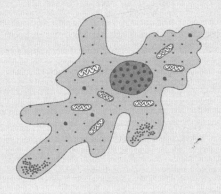

BY ANNE MACZULAK

SECOND EDITION

The Smart Guide To Biology - Second Edition

Published by

Smart Guide Publications, Inc.
2517 Deer Chase Drive
Norman, OK 73071
www.smartguidepublications.com

For information, address: Smart Guide Publications, Inc. 2517 Deer Creek Drive, Norman, OK 73071

International Standard Book Number: 978-1-937636-81-4

Library of Congress Catalog Card Number:
11 12 13 14 15   10 9 8 7 6 5 4 3 2 1

Printed in the United States of America

Cover design: Lorna Llewellyn
Copy Editor: Ruth Strother
Back cover design: Joel Friedlander, Eric Gelb, Deon Seifert
Back cover copy: Eric Gelb, Deon Seifert
Illustrations: Lorna Llewellyn
Production: Zoë Lonergan
Indexer: Cory Emberson
V.P./Business Manager: Cathy Barker

ACKNOWLEDGMENTS

My appreciation goes to Smart Guide Publications and my agent Jodie Rhodes for developing the idea for this book and the superb editorial help of Ruth Strother who did an extraordinary job editing this book. I also thank illustrator Lorna Llewellyn who did a wonderful job illustrating the book.

Last but not least, I owe special gratitude to my writers group. We have supported and encouraged each other for many years. My writing career would not have begun without the inspiration I received from Sheldon Siegel, Priscilla Royal, Bonnie DeClark, Meg Stiefvater, and Janet Wallace.

TABLE OF CONTENTS

INTRODUCTION. xv

PART ONE: *Cells* 1

1 The Cell. 3
Our Common Ancestor 3
Chemicals to Cells 5
The Two Types of Cells 6
Basic Cell Structure 7
   Membranes. 7
   Cytoplasm 8
   Genetic Material. 9

2 How The First Cells Emerged on Earth. 11
Carbon 12
   The First Organic Molecules 14
How Chemicals Lead to Life 15
Amino Acids and Proteins 16
   Protein Structure 17
Ribonucleic Acid (RNA). 17
Deoxyribonucleic Acid (DNA) 18
The First Cell 18

3 The Prokaryotic Cell 19
Bacteria and Archaea 19
Inside Prokaryotes 21
Outside Prokaryotes 21
Diversity in Prokaryotes. 23
   Extremophiles 24
   Cyanobacteria 25
The Domains of Living Things 26

4 The Eukaryotic Cell 27
The First Eukaryotes. 28
What Is an Organelle? 28
Protists 30
   Algae 31
   Diatoms 33

Water Molds, White Rusts, and Downy Mildews 33
Protozoa 34
Algae Lead to the First Green Plants 34

5 *How Your Cells Work* 37

Biological Membranes 38
Lipid Bilayer 39
How Substances Go through Membranes 40
Energy Generation in Membranes 40
How Membranes Help Cells Communicate 42
Mitochondria 42
The Nucleus 43
Endoplasmic Reticulum (ER) 43
Golgi Apparatus 44
Other Parts of Your Cells 44
Peroxisomes 44
Lysosomes 45

6 *How Cells Communicate* 47

Chemotaxis 48
Quorum Sensing 49
Multicellular Organisms 50
An Early Multicellular Organism? 50
Cell Signaling 51
Neurons 52
Controlling Gene Expression 53

PART TWO: *Energy* 55

7 *Introduction to Metabolism* 57

Enzymes 57
How Enzymes Run Chemical Reactions 58
Coenzymes 59
Regulating Enzymes 59
Anabolism and Catabolism 61
Equilibrium 62
ATP - The Cell's Energy Currency 62
How ATP Stores Energy 63
How ATP Performs Work 63
How ATP Regenerates 63

8 Respiration 65

How We Get Energy from Organic Fuel 65
Glycolysis 66
Krebs Cycle = Citric Acid Cycle = Tricarboxylic Acid Cycle 67
Electron Transport 68
Phosphorylation - Big Word with a Big Meaning 69
Oxygen 70
Respiration's ATP Output 70

9 Fermentation 71

Getting Energy without Using Oxygen 72
Types of Fermentation 74
Alcohol Fermentation 75
Lactic Acid Fermentation 75
Mixed Acid Fermentation 76

10 Photosynthesis 77

Photosynthetic Organisms 78
Cyanobacteria 79
Chloroplasts 80
Light and Photosynthetic Pigments 81
Photosynthesis in Bacteria 82
Photosynthesis' Two Stages 82
Light Reactions 83
Dark Reactions 84

PART THREE: *Basics of Genetics* 87

11 Deoxyribonucleic Acid (DNA) and Chromosomes 89

Your Cells' DNA 89
Complementary Strands 91
The Genetic Alphabet 91
What Is a Gene? 91
How DNA Replicates 93
Chromosomes 94
How Chromosomes Control the Traits You Inherit 95

12 The Cell Cycle 99

Cell Division 99

Cell Division in Prokaryotes 100
Cell Division in Eukaryotes 100
Mitosis. 100
The Importance of Cell Cycles 102
A Typical Cell Cycle 102
How Nature Controls Cell Cycles 102
Cancer: When the Cell Cycle Goes Out of Control 103

13 Genetic Information Travels from Genes to Proteins 105

What Does Gene Expression Mean? 105
RNA's Job in Transcription 106
RNA's Job in Translation 107
The Genetic Code 107
Other Components of Gene Expression 108
Protein: The Final Product 109

14 Genetics and Inheritance 111

Heredity, Genetics, and Inheritance 112
Mendelian Genetics 112
Law of Segregation. 114
Law of Independent Assortment 114
Following Traits through Generations 115
Recessive and Dominant Traits 116
How Genes Are Linked on Chromosomes 117
Sex-Linked Genes 118

PART FOUR: *Adaptation and Evolution* 121

15 Mutation and Other Genetic Errors 123

What Is Mutation? 123
Types of Mutations. 124
What Causes Mutations? 124
Mutation Rate 125
How Mutation Leads to Adaptation 125
An Introduction to Natural Selection .127

16 Evolution and Extinction 129

Even the Theory of Evolution Evolved! 129
Catastrophism 130
Gradualism 131

Lamarck's Ideas on Evolution 131
Darwin's Theory of Evolution 132
The Origin of Species 132
The Relationship between Evolution and Biodiversity 134
Extinction 134
Types of Extinction 135

17 How We Organize Species 137

Phylogeny, Systematics, and Taxonomy! Oh My! 137
How Hierarchies Work 138
Domains 139
Kingdoms 140
A Taxonomist's Life 140
How Humans Fit into Taxonomy 141

PART FIVE: *Prokaryotes, Eukaryotes, and Viruses* 145

18 Single-Celled Organisms: Bacteria and Archaea 147

Archaea 148
Sulfur Users 148
Methane Producers 148
Halophiles 149
Archaea without Cell Walls 149
Bacteria 150
Photosynthetic Bacteria 152
Proteobacteria 152
Spirochetes 153
Purple and Green Sulfur and Nonsulfur Bacteria 153
Mycobacteria 153
Mycoplasma 154

19 Algae, Protists, and Fungi 155

Algae 156
Protists 157
Slime Molds 159
Fungi 160
Multicellular Fungi 160
Mushrooms 162
Molds 162
Yeasts 163

20 Viviruses 165

Meet a Virus 166
  Bacteriophages: Viruses that Attack Bacteria 166
  Virus Size 167
How We Classify Viruses 167
How Viruses Infect 168
  The Lytic Cycle of Virus Infection .169
  The Lysogenic Cycle of Virus Infection 169
Viral Diseases 170
Other Nonliving Particles in Biology .170
  Viroids 170
  Prions 170

PART SIX: *Complex Animals* 173

21 Invertebrates 175

Characteristics of Early Organisms 176
  Sponges: Exceptions to the Animal Body Plan 176
Invertebrates That Do Not Look Like Animals 177
Invertebrates That Look Like Animals .178
  Chordata 180

22 The Animal Body: The Nervous System and Circulation 183

Tissue Structure and Function 184
  Epithelial Tissue 184
  Muscle Tissue 185
  Connective Tissue 186
  Nerve Tissue 186
Body Metabolism: How It All Works Together 187
  Ectotherms and Endotherms 187
  Thermoregulation 188
The Nervous System 189
  Neurons and Glia 189
The Circulatory System 189
  Open and Closed Circulatory Systems 190
  Blood Vessels 191
  Blood 191

23 The Animal Body: The Endocrine, Immune, and Sensory Systems 193

The Endocrine System. 194
Endocrine Glands 194
Hormones 195
Pheromones 195
The Immune System 196
Components of the Immune System 196
Cells of the Immune System. 198
Antibodies 198
Natural Immunity 198
Acquired Immunity 198
The Sensory System 199
Types of Sensory Receptors 199
Components of the Sensory System. 200

24 The Basics of Animal Reproduction. 201

Reproductive Cycles 202
Asexual Reproduction in Animals 202
Fertilization. 203
Parts of the Female Reproductive System 203
Parts of the Male Reproductive System 204
Breeding. 205

PART SEVEN: *Plants* 209

25 The Plant Body and Plant Growth 211

Plant Diversity 211
Early Plant Life 212
Mosses 213
Ferns 213
Vascular Plants 214
Anatomy of a Plant. 215
Roots 215
Stems 216
Leaves. 217

26 Seeds and Plant Life Cycles. 219

What Is a Plant Life Cycle? 219
Seeds. 221

Why Seeds Are Important to Us 222
Life Cycle of Seed Plants. 223

27 Flowers, Fruits, and Pollination 225

Parts of a Flower 225
   Carpels 226
   The Stigma 226
   Stamens 227
   Other Parts of a Flower 227
Classifying Flowers. 228
Pollination 228
Insects and Pollination. 229
Fruits 229
An Angiosperm Life Cycle 230

28 Plant Sensory and Defense Systems 233

Nature's Stimuli Sensed by Plants. 234
   A Plant's Signal Response System. 235
Plant Hormones 236
   Fruit Growth and Ripening 237
Types of Plant Defenses 238
   Plant Defenses against Herbivores 239
   Plant Defenses against Pathogens 239

PART EIGHT: *Environment and Ecology* 243

29 Environments 245

The Earth's Environments. 246
   Soil Environments 248
   Water Environments 248
Ecosystems 250

30 Ecology 253

How Humans Fit into Earth Ecology 253
Human Activities and Ecology 254
How Species Interact. 255
Nutrients Cycle through Earth's Ecosystems 256
Prey and Predators. 257
Biodiversity 258
   Biodiversity and Ecosystems 258

Climate and Biodiversity 259
Deforestation and Desertification 259

31 Biodiversity 261

The World's Biodiversity Hotspots 262
Endemism 263
Ways of Protecting Biodiversity. 263
Threats to Biodiversity. 263
The Value of Biodiversity 265
Indicator Species. 265

32 Today's Biology 267

Different Types of Biology. 268
Molecular Biology 268
Genomics 269
Proteomics 270
Conservation Biology 272

33 The New Frontiers In Disease Fighting 273

How New Technologies Are Developed 274
Biotechnology 276
Today's New Technologies in Disease Fighting 277
Biologics. 279
Gene Therapy. 280
Which diseases will new technologies target first? .281

34 How To Understand Biology In The News 283

How Scientists Talk 285
Knowing a Little Statistics Can't Hurt 286
How to Spot Junk Science. 288
Our Perception of Biology. 291

Appendix A - Glossary 293

Index 299

About The Author 318

INTRODUCTION

Biology is the study of all living things. That's a pretty big subject. This Smart Guide breaks biology down into easy-to-manage pieces. It starts with the basic unit of all living things, the cell. After taking a tour of the basic types of cells that make up all organisms, you embark on a voyage through biology.

This book covers cell communication, how cells developed on Earth, and how ancient cells evolved into today's organisms. You then learn about the inner workings of modern organisms. This book introduces you to DNA and proteins, genetics and inheritance, and bacteria, algae, plants, animals, and every other living thing.

You will meet the invertebrates here as well as more complex animals and the Earth's plants. The organs and systems that make your body operate are described in this Smart Guide. Finally, this Smart Guide shows you the big picture of biology. You will gain an understanding of ecology and biodiversity. This book concludes with an overview of the new technologies that biologists now use to study the living world.

The *Smart Guide to Biology* gives you a handy resource to the basics of biology.

PART ONE

Cells

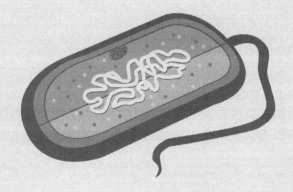

CHAPTER 1

 **The Cell**

In This Chapter

➤ Relating cells to our common ancestor in evolution

➤ The chemical composition of biological cells

➤ Two types of cells in nature: prokaryotes and eukaryotes

➤ The basic components of all living cells

In this chapter you'll discover the most basic unit of all living things: the cell. A cell is the simplest collection of matter that can live. All life begins today and all of Earth's life started almost four billion years ago with a single cell.

Scientists have pointed out that the cell is as fundamental to biology as the atom is to chemistry. Since biology is the study of living things, we must begin where biology began.

Our Common Ancestor

Throughout this book you will see how living things, called *biota*, relate to each other. This relationship starts with cells. All biota are made of up cells. Whether you are now envisioning a rhinoceros, an elm tree, a *bacterium*, or the human body, they all have in common a structure built on tiny cells.

Most plant and animal cells have diameters from 10 micrometers to 100 micrometers, or about 0.0004 to 0.004 inch. Bacterial cells measure a bit smaller, from about one micrometer to 10 micrometers, but some of the largest bacteria are bigger than the smallest plant or animal cells. This book shows you the ways in which biology offers many common themes for us to consider when we think about nature. But simultaneously, there often seems to be little consistency across nature. As a case in point, the largest bacteria can be seen without a microscope; the smallest invertebrates can be spotted only by using a microscope.

A blue whale is Earth's largest mammal, 1,600 times heavier than a 175-pound human. Yet a whale's cells are not the size of beach balls but rather the same size as the cells that make up a human. This and other features of cells give testament to the steadfastness of biology. Biology follows certain rules yet somehow manages to give rise to staggering diversity.

Diversity in Earth's biology comes from a common ancestor. This common ancestor is the theoretical first cell that developed 3.5 billion years ago. I call it a "theoretical" cell because the first cells probably evolved in many similar events in different places on the young Earth and at different times. Evolution may be easier to understand, however, by thinking about a single cell from which all other life originated.

Biology Tidbits

The Cell Theory

The idea that all living things are composed of one or many basic subunits, or cells, is called the cell theory. The English inventor Robert Hooke first coined the term "cells" in 1665 when observing various snippets of nature in his homemade microscope. By observing the microscopic structure of leaves, cork, insects, and other matter plucked from nature, Hooke was the first to realize that all life might be defined by this single, simple building block.

The vast diversity of life we know today could well have evolved into an equally diverse collection of cell types. Instead, all Earth's biota contains only one of two different cell types. That's it! All of biology comes down to two different types of cells … and one of the types, *eukaryotes*, evolved from the first type, *prokaryotes*. Evolution can be a complex subject, and we will revisit it several times in this book, but the best place to start learning about biology and evolution is by knowing something about our common ancestor: the prokaryotic cell.

Chemicals to Cells

A cell is the smallest unit of life. Some cells live completely independent lives and other cells must live in communities. Nevertheless, every cell has some independence. A cell gets its own food. It has a mechanism to turn that food into energy that enables it to get more food, reproduce, and do work. (A cell's work amounts to sensing conditions in the environment, making adjustments to changes in the environment, and sometimes swimming.)

Biology Tidbits

Atoms and Elements

Let's take a short chemistry lesson to see how chemicals relate to life. The building block of all matter, living or nonliving, is the atom. An atom is a submicroscopic particle, meaning it is too small to be seen with conventional microscopes. Atoms contain a central area called the nucleus made up of positively charged *protons* and neutral *neutrons*. Negatively charged *electrons* orbit the nucleus like planets orbit the Sun. Therefore, an atom is the sum of protons, neutrons, and electrons, and this combination is indivisible; nature cannot divide an atom.

An element is matter in its simplest form. Put another way, an element consists only of atoms of the same type. The element gold contains only gold atoms; the element nickel contains only nickel atoms; carbon contains only carbon atoms, and so on.

What makes carbon different from gold or from any other element such as hydrogen, fluorine, or plutonium? The structure of their atoms determine the element. The atoms that make up carbon contain six protons and six neutrons in their nucleus. No other element contains this configuration. By comparison, gold has 79 protons and 118 neutrons; phosphorus has 15 protons and 16 neutrons. Find a periodic table of chemical elements to see all the other elements that make up living and nonliving things on Earth.

Before cells could assemble into living things, they first were nothing more than collections of chemicals with no function and no life. All cells evolved using the chemicals available on the young Earth. Today's cells reflect the elements available billions of years ago. Thus, hydrogen, nitrogen, oxygen, carbon, sulfur, and phosphorus make up more than 99 percent of all cells.

As we build a cell throughout the early chapters of this book, you will see how these elements come together to form *proteins*, carbohydrates, fats, and other large molecules that make up the actual structure of a cell.

Biology Vocabulary

Molecules and Compounds

Biologists view the world more in terms of molecules and compounds than atoms and elements. Biota are defined by the molecules and compounds from which they are built.

A molecule is a chemical union of two or more atoms of the same element or different elements. For example, a molecule of hydrogen is H2, meaning it contains two hydrogen (H) atoms, but a molecule of water is H2O, meaning it contains two hydrogens and one oxygen (O) atom. A compound is a chemical union of two or more different elements. Therefore, water is a compound of hydrogen and oxygen. Glucose (C_6H12O_6) is a compound of carbon (C), hydrogen, and oxygen.

The Two Types of Cells

Implausible as it seems, the Earth holds only two types of cells. These are prokaryotes and eukaryotes.

Prokaryotes are the most numerous and diverse of all biota. They are also invisible! Except for a few rare exceptions, prokaryotes cannot be seen without a microscope. They number in the millions of billions on Earth, too numerous to estimate their numbers with any trusted accuracy. Scientists have proposed the numbers of bacteria on Earth as being more than 10^{30}. To visualize this number, write the numeral "1" and then put 30 zeros behind it.

Although we tend to use the terms "prokaryote" and "bacteria" interchangeably, a second type of cell also belongs with the prokaryotes. These cells are called *archaea*. Archaea are similar to bacteria and under a microscope you would not see any difference. Archaea differ slightly from bacteria in their composition and in the places where they live on Earth.

In Chapter 18, you will see how the habitats of archaea relate to the earliest conditions on Earth and to evolution.

Basic Cell Structure

All cells follow a common theme. You can think of any cell as a sac of water. The sac must have some sort of outer boundry to keep the water from escaping. The watery contents must also contain compounds that enable the cell to get around, replicate, and carry out its basic functions.

Biologists have given names to these three components. They are *membranes*, the *cytoplasm*, and the cell's genetic material.

Membranes

Membranes are thin layers of proteins and fatty substances called *lipids*. All biological membranes separate the watery interior of a cell from the exterior of the environment. Usually, the environment is also watery.

Why didn't nature invent a barrier like aluminum foil or a brick wall to keep the cell's insides from mixing with stuff on the outside? Nature invented membranes because these barriers serve an important role: They can choose what substances get into the cell and which substances exit the cell. This ability has the fancy name "selective permeability." Anything that is *permeable* has the ability to allow certain substances to pass through it.

Biological membranes have specific layers inside them. The membrane's proteins and lipids align in a way to both protect the cell's contents and also give the cell the ability to choose the materials it takes in and sends out. (I say "biological membranes" because scientists have invented nonliving membranes for a variety of other purposes. The purified water used in making semiconductors has been first passed through a membrane. The biological membranes of cells are sometimes called "plasma membranes" to distinguish them from nonliving membranes.)

Biology Vocabulary

Macromolecules

Macromolecules are large organic compounds. Although a molecule does not need to exceed a particular size to be called a "macromolecule," biologists usually reserve this term for substances such as proteins, long-chain carbohydrates called *polysaccharides*, large lipids such as *sterols*, and deoxyribonucleic acid (DNA).

A membrane's layers have been called a "sandwich" because from the outside to the inside the composition is: hydrophilic (water-attracting) lipid, hydrophobic (water-repelling) lipid, and hydrophilic lipid. Thus, the interior of the membrane is incompatible with water soluble compounds and so is the outside of the membrane. Since many nutrients a cell needs are water soluble, how does the cell get these things across its membrane if the membrane interior repels water? Proteins interspersed in the membrane play an important part in helping to transport nutrients into the cell. These proteins dot the membrane like raisins in an oatmeal cookie; some proteins extend completely across the membrane and others go only halfway.

The membrane is flexible and somewhat fragile. To help strengthen a cell's resistance to a harsh environment, plant cells and prokaryotes have an exterior cell wall that lies on top of the membrane. Animal cells lack this protection.

Cytoplasm

The cell interior is called the cytoplasm. It is about 80 percent water. The remainder is made up of the cell constituents that make the cell function. Thus, the activities concerned with nutrient use, energy generation, waste consolidation, movement, and replication take place in whole or in part in the cytoplasm.

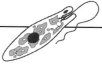

Inside a Cell

Water and Life

About 99 percent of a cell's molecules are water molecules. By weight, the water makes up about 70 percent of the cell. So vital is water to your life, you may actually take it for granted and forget that it is your most important nutrient.

Water offers cells a place where chemical reactions can occur. Without this matrix, the rest of a cell's systems would stop. Water molecules also participate in many reactions. The H_2O donates hydrogen or oxygen, or both, in certain chemical reactions that turn one compound into a different compound.

In a microscope, the cytoplasm of prokaryotes looks simpler than that of eukaryotes. Prokaryotic cytoplasm contains fewer structures. It contains a general area where the prokaryote's genetic material lies and also tiny dot-like particles called *ribosomes*. Ribosomes act as the place where the cell makes new proteins. Some prokaryotes have additional structures scattered in their cytoplasm, but their contents remain relatively simple and homogeneous.

Eukaryotic cytoplasm is stuffed with more defined structures. These subcellular structures are called organelles and you will learn about the main ones in chapter 5. One important point about eukaryotic cells compared with prokaryotic ones is: eukaryotes contain membrane-bound organelles. Except for ribosomes, the structures inside eukaryotic cells are enclosed in another membrane. Therefore, eukaryotes have an outer plasma membrane and also smaller membranes for each individual organelle.

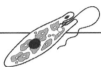

Inside a Cell

To Membrane or Not to Membrane?

Eukaryotic cells seem to have an advantage over prokaryotes because they protect each of their internal structures from physical harm by enclosing it with an additional membrane. Prokaryotes don't mess around with extra membranes; they let their internal structures float free in their cytoplasm. But biologists believe that the simple architecture of prokaryotes give them the advantage.

First, it takes extra energy for eukaryotic cells to move substances back and forth across all those membranes. Prokaryotes save this energy for other things. Second, prokaryotes have an easier time sensing conditions in their surroundings and getting that information to mechanisms that help the cell react. Eukaryotes, by contrast, are less efficient at this. Third, prokaryotes can be more compact than eukaryotes, which need extra room for their specialized membrane-bound compartments. The compact structure of prokaryotes increases their surface area-to-volume ratio. The high surface-to-volume ratio further enhances a prokaryote's ability to "read" its surroundings and respond to danger quicker than eukaryotes.

Genetic Material

All the genetic material of an organism is collectively known as its *chromosome*. The chromosome carries all the information about you as a member of your species, specific information that makes you a unique individual, and the information that you will pass to your offspring.

In prokaryotes, the chromosome floats in the cytoplasm in two places. The main body of chromosome exists as a highly coiled strand of the macromolecule deoxyribonucleic acid (DNA). Smaller bits of DNA also float in the cytoplasm separate from the main chromosome. These small pieces are called *plasmids*. Plasmids are also made of DNA and usually have a circular shape so that the DNA strand has no beginning and no end.

In eukaryotes, the DNA lies inside an organelle called the *nucleus*. (This is different from the nucleus of an atom.) Eukaryotic cells protect their DNA by enclosing the nucleus in a double membrane layer. Between the two membranes lies a net-like matrix that gives the nucleus shape and provides some strength.

Eukaryotes also organize their DNA differently than prokaryotic cells. The DNA is organized into discrete areas, each of which is called a chromosome. For this reason, we think of eukaryotes as having multiple chromosomes but bacteria contain just one chromosome. All animals have a characteristic number of chromosomes. For instance, humans contain 46 chromosomes, ants have two, and a crayfish has 200.

While prokaryotes compact their chromosome by coiling it into a dense ball, eukaryotes wrap their chromosomes around proteins called histones. Either type of organization represents a key housekeeping priority for the cell. The DNA is so massive a molecule, if unraveled it would stretch a thousand times the diameter of the cell. Yet to reproduce, a cell must also replicate its chromosome. Cells do this by replicating sections of DNA at a time rather than unspooling the entire DNA molecule.

The organization, management, and replication of the chromosome represent one of the most complicated tasks of any cell.

CHAPTER 2

How The First Cells Emerged on Earth

In This Chapter

➤ Introduction to the chemicals that led to the first cells

➤ Earth's first organic compounds and how they react

➤ The basics of amino acids, proteins, and nucleic acids

➤ How the first cell likely formed

In this chapter you will discover how the first cells formed on Earth. What extraordinary events must have occurred for a mish-mash of chemicals to align in a way that allowed life to emerge? Even more fascinating to biologists is that life formed under the harshest of conditions.

The Earth formed about 10 billion years ago as a molten ball of metal shucked off from the Sun. The Sun had been produced from the same helium and hydrogen that formed the rest of our solar system. This star's heat was hot enough to fuse hydrogen and helium atoms repeatedly, forming all of the other elements that now make up the Earth.

After 5.5 billion years, the planet had cooled enough to allow chemical *reactions* to occur with a modicum of order rather than the random and violent collisions of atoms that had been occurring previously. But the Earth was still a violent, boiling, and caustic place. Volcanic eruptions happened everywhere and electric storms filled the sky. The air lacked oxygen; the atmosphere was mainly hydrogen, *methane*, and *ammonia*.

Could life start under these conditions? Some people believe evolution is impossible mainly because of the improbabilities that would have had to occur. But biology is not like chemistry or physics. In those sciences, the impossible never happens. Sugar and water never mix to form gold; rocks never tumble off a mountain and fly into the sky. In biology, however, the seemingly impossible can happen. While biology follows certain rules, it also modifies those rules with regularity. Somehow, the implausible occurred and life emerged on a lifeless planet.

This chapter introduces you to the chemicals that merged to form other chemicals leading to the first cells. This fantastic journey had no human witnesses. No records exist pointing to the first cell. Only modern *biochemistry* has demonstrated the steps taken from simple chemicals to more complex compounds to the macromolecules that form a living thing. The story begins with the element carbon.

Biology Tidbits

The First Spark of Life

In 1953, Harold Urey and Stanley Miller set up a flask filled with the gases known to exist in early Earth's atmosphere. They boiled water in a separate connecting flask to add vapor to the mix, and then passed an electrical charge through the milieu. After a week, Urey and Miller tested the compounds in the flask. They discovered various amino acids, sugars, fats, and more complex organic compounds, such as *nucleic acids*. The Urey-Miller experiments showed that the conditions on Earth 4.5 billion years ago could give rise to the compounds that would be the precursors to life.

Carbon

The element central to all life makes up only 0.09 percent of the Earth's crust. An organism could not exist without carbon. Every cellular process possessed by you, by bacteria, by plants, and every other living thing starts with carbon.

Carbon plays a useful role in many different types of compounds mainly because a carbon atom can form chemical bonds with up to four other atoms at once. For example, the natural gas methane (CH_4) contains one carbon connected to four hydrogens. Carbon can also form a double set of connections with itself, called a double bond. The gas carbon dioxide (CO_2) contains such a double bond. I could also write carbon dioxide as "O-C=C-O."

Scientists call carbon-containing compounds organic compounds, and there are more than five million of them on Earth. Carbon forms the backbone for carbohydrates that give the body energy and store energy. For example, the sugar glucose is an essential carbohydrate for cells. For storing excess glucose, the liver strings many glucose molecules together to form the macromolecule *glycogen*. In this way, glycogen stores energy for the body's later use.

Carbon also provides the backbone for the types of compounds that represent life: proteins, *enzymes*, lipids, fats, and *polymers*, which are any long compounds made up many subunits. Carbon is essential in forming carbon dioxide, without which photosynthesis would not occur. Without photosynthesis replenishing the atmosphere with oxygen, you would not survive!

Carbon also plays an integral part in hydrocarbons. These are long strands of carbon linked to each other and each also carrying hydrogen atoms. Hydrocarbons form coal and oil. We call these hydrocarbons "fossil fuels" because they actually contain fossils of primitive biota that lived on Earth about 300 million years ago.

The first cells emerged in Earth's oceans. Over billions of years they merged to form more complex organisms. These tiny primitive organisms dominated ocean life, and as each generation died they sank to the depths. (The majority of the world's carbon still resides in our oceans.) The weight of water and sediment pressed the dead organisms into the seabed. The immense weight plus the nature of the carbon-hydrogen bond forced water to be expelled from the carbon-rich layer. Over millions of years, the hydrocarbons from ancient organisms solidified into a dark mass that we today recognize as coal. The deepest layers migrated toward the hot center of the planet and changed into a thick liquid. This liquid we know as petroleum or oil.

Biology Vocabulary

Carbohydrates and Hydrocarbons

Carbohydrates are sugars, such as glucose and fructose, and long chains of sugars. The long chains are called polysaccharides, but they too are carbohydrates.

Hydrocarbons differ from carbohydrates by a lack of oxygen. Some hydrocarbons are immense and composed of 30 or more carbons each carrying two hydrogens. Part of a hydrocarbon chain looks like this:....-CH_2-CH_2-CH_2-CH_2-CH_2-CH_2-CH_2-....

Even before carbon began its transformation to fossil fuels belowground, it had served as the starting point for the first organic molecules aboveground. This would be the first step toward life.

The First Organic Molecules

The atmosphere on Earth 4.5 billion years ago contained mainly the gases hydrogen, methane, and ammonia with bursts of water vapor liberated from molten rocks and volcanoes. Lightning pulsed through this atmosphere repeatedly because the young Earth endured almost constant violent storms. The energy and heat supplied by the lightning caused an occasional merging of the carbon, hydrogen, nitrogen, and oxygen in the atmosphere's gases and vapors. As a result, amino acids formed.

Biology Vocabulary

Amino Acids

Amino acids are the building blocks of proteins. A protein is simply a strand of amino acids linked together and then folded up in a particular shape to help the protein do its job in a cell. The folding and the protein's function depend on a particular order of amino acids in the strand. Take one amino acid out and replace it with the wrong one and a protein might be rendered useless to a cell.

On ancient Earth, carbon, hydrogen, nitrogen, and oxygen assembled into various chemical arrangements. Intense heat at the Earth's surface and electrical storms provided the energy for these chemical couplings. Certain configurations formed more easily than others, so particular compounds began to accumulate. Today, biologists recognize six basic molecules that give rise to almost all cellular structure. We can think of these compounds as our link to the Earth's first organic molecules:

➤ Amino acids—The building blocks of proteins, amino acids always contain an amino portion (NH_2) linked to a carbon and a carboxyl portion (COOH).

➤ Glycerol—A three-carbon compound in which each carbon also connects to a hydroxyl (OH). Glycerol forms the backbone of many complex lipids.

➤ Sugars—A major source of energy for cells, these simple carbohydrates usually contain five or six carbons connected to each other and also to hydrogen and oxygen.

➤ Fatty acids—Another source of energy, these are hydrocarbons that always have a carboxyl (COOH) at one end.

➤ Phosphate—This is the first of the "ancient molecules" in which a new element appears. Phosphates contain phosphorus connected to oxygen atoms and are essential in transferring energy from one compound to another compound inside cells.

➤ Nucleotides—The most complex of the ancient molecules, nucleotides are the backbone of nucleic acids, the genetic material of all biota. Nucleotides contain a nitrogen-containing portion, a sugar, and a phosphate.

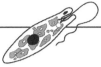

Inside a Cell

Nitrogen

Nitrogen gas (N2) makes up about 80 percent of the atmosphere, but living things have a hard time getting nitrogen into their bodies. Animals cannot inhale nitrogen as they inhale oxygen; plants cannot absorb nitrogen in the same way they absorb carbon dioxide. Yet nitrogen is essential in amino acids and thus proteins. It is also an essential part of nucleic acids and vitamins. Energy processes inside cells could not occur without nitrogen. This is because nitrogen plays a role in the energy-generating steps that occur in membranes, and the nitrogen is a component of the membrane's structure as well.

How Chemicals Lead to Life

After the amino acids, sugars, and other key components of life began to accumulate on Earth, did life miraculously spring forth? The main transition from lifeless chemicals to a cell occurred when certain elements formed *functional groups* on the organic compounds already present. A functional group is a cluster of atoms attached to a compound that enables the compound to react with others.

A phosphate group containing a phosphorus atom connected to four oxygen atoms gives an example of a functional group. The main functional groups in biology are:

➤ Phosphate ($-PO_4$)—It is involved in energy transfer reactions. (The dash in this abbreviation and the others below indicates that the functional group attaches to some other compound.

➤ Hydroxyl (-OH)—It collects electrons in certain reactions and attracts water molecules to help dissolve substances such as sugars.

➤ Carboxyl (-COOH)—Acts as an acid by donating its hydrogen atom in certain cellular reactions.

> ➤ Amino ($-NH_2$)—Acts as a base by picking up excess hydrogens.

> ➤ Sulfhydryl (-SH)—Works mainly to create bridges between different parts of the same protein, thereby stabilizing the protein.

All of the above functional groups offer another benefit; they help a compound dissolve in water. Since the first cells likely began in the oceans, water solubility is a definite plus in helping reactions go forward.

Biology Vocabulary

Lipids and Fats

Lipid is the name for a group of diverse water insoluble compounds. Lipids include fats and sterols. (Sterols serve the body in one way by being a precursor for making vitamin D.)

Fats are organic compounds that have glycerol as a backbone. A glycerol molecule can hold up to three chains called fatty acids. Fatty acids are similar to hydrocarbons because they don't get along well with water. The difference is that fatty acids contain oxygen and hydrocarbons do not.

Amino Acids and Proteins

Amino acids possess both an amino group and a carboxyl group connected to each side of a central carbon. Because amino acids carry these two functional groups, they become important in cellular reactions because they can act as an acid and a base. Very few other compounds in nature have this talent.

The amino and carboxyl groups account for two of the four spots on carbon. What fills the remaining two spots? One location is always taken by a hydrogen atom. The fourth location goes to a variety of chemical structures that give each amino acid in biology a unique look. This variable sidekick is sometimes referred to as an R group, where "R" stands for any of the diverse structures in nature that can be hooked onto an amino acid.

Nature has invented more than 500 different amino acids, but animal cells use only 20 to make all the proteins they need. The 20 amino acids used in making animal protein are: alanine, arginine, asparagine, aspartic acid, cysteine, glutamic acid, glutamine, glycine, histidine, isoleucine, leucine, lysine, methionine, phenylalanine, proline, serine, threonine, tryptophan, tyrosine, and valine.

The human body can make 10 amino acids for itself but must have the remaining 10 supplied in the diet. These dietary amino acids are called the essential amino acids.

Biology Vocabulary

Essential Amino Acids

Every animal cell has a requirement for certain amino acids that must be supplied from the outside. In people, these amino acids come from our diet. They are therefore defined as nutrients and are called essential amino acids. The essential amino acids are: arginine, histidine, isoleucine, leucine, lysine, methionine, phenylalanine, threonine, tryptophan, and valine. Usually, youngsters require arginine in their diet but this dietary requirement fades as a person ages.

Since prokaryotes and plants do not have a diet, they must make all of the amino acids they need for protein building.

Protein Structure

Cells must assemble their proteins in a specific way for the protein to work as it should. Every protein has a primary, secondary, and tertiary structure from which a cell must not deviate. Certain specialized proteins have an added quaternary structure. All of this sounds very complicated, but put simply, it means the following:

➤ Primary—The specific order of amino acids in the protein.

➤ Secondary—The first set of folds that partially compacts a protein.

➤ Tertiary—The second set of folds that bends the protein back on itself and makes it compact.

➤ Quaternary—A situation in which two or more proteins attach to each other to form a mega-protein!

Ribonucleic Acid (RNA)

Between 4.5 billion and 3.5 billion years ago, the first primitive molecules became more complex and then a few connected to each other to form bigger molecules. When nucleotides linked up this way, they began making the earliest ribonucleic acid or RNA.

RNA consists of a string of the same sugar called ribose linked by a phosphate. Each ribose holds onto a nitrogen-containing compound called a base. (The sugar plus phosphate plus base equals a nucleotide.)

The formation of RNA would be a pivotal point in cell evolution because nucleic acids, such as RNA, carry information that goes to new cells when they replicate.

Deoxyribonucleic Acid (DNA)

You may know deoxyribonucleic acid, or DNA, as the main stuff that makes up your chromosome. Nonetheless, this nucleic acid probably developed after RNA. DNA resembles RNA in structure. The main differences between these two macromolecules is: (1) DNA contains slightly different bases than RNA and (2) DNA consists of two chains linked together via their bases.

The First Cell

A monumental leap occurred on Earth sometime between 4.0 billion and 3.8 billion years ago. During that period, storms became less tumultuous and a bit more sunlight broke through the atmosphere. Amino acids began linking together to form short chains (called peptides). Nucleotides began forming, and these also linked up to form chains.

In a watery environment, fatty compounds preferred not to float free since fats and water repel each other. The fats spontaneously ganged up to form little hollow balls in the ocean. By balling up, they reduced their exposure to water. As a result, rudimentary membranes developed, holding the ocean out of the center of the ball. The center of each ball contained a tiny volume of ocean water. Occasionally, the creation of a fatty ball also included one or more peptides and nucleotide chains.

Any membrane-enclosed ball that contained both peptides and primitive RNA held the potential to do two crucial things. First, the peptides could lead to proteins, and these would give the cell the possibility of carrying out functions. Second, RNA had the ability to self-replicate. Self-replication would turn a glob of primordial chemicals into an operational, reproducing cell!

From this point onward, cells might add new capabilities, learn to sense their environment, and team up to create multi-celled organisms.

The time needed to progress from a newly born Earth to the first precursor of a cell took about 6.5 billion years. All the additional life that you know today needed only 3.5 billion additional years to develop!

CHAPTER 3

The Prokaryotic Cell

In This Chapter

➤ The differences and similarities of bacteria and archaea

➤ The features inside prokaryotic cells

➤ The features on the outside of prokaryotic cells

➤ Characteristics of extremophiles and cyanobacteria

In this chapter you will learn about the simplest of all living things. Although prokaryotes are the smallest and simplest cells that can live independently, they also carry enormous responsibilities in keeping all other organisms on Earth alive. In later chapters, you will see how prokaryotes recycle the Earth's nutrients, decompose waste, and make the Sun's energy available to all animal life.

Let's begin with the cells from which humans and every other complex organism evolved.

Bacteria and Archaea

You may think of bacteria as microscopic specks that live almost everywhere, or as germs that cause disease, or as allies who help us digest our food and degrade our wastes. Bacteria do all of these things and more.

A lesser known group of prokaryotes, the archaea, handle equally important roles on Earth. Archaea tend to live in remote places where humans cannot or choose not to go. Although archaea look just like bacteria in a microscope, they possess features that help them withstand the harsh environments where they sometimes live.

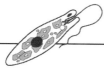

Inside a Cell

Oxygen: The Most Toxic Element

Life emerged on Earth at a time when oxygen did not fill the atmosphere. Only trace amounts of oxygen occurred four million years ago and that oxygen was bound to water vapor rather than the gas we breathe today.

Life evolved in an oxygenless or anaerobic environment. Any free oxygen atoms that entered the first cells were toxic to them. This is because oxygen reacts with other molecules to form unstable chemicals called free radicals. Free radicals are as dangerous to cells today as they were to the first cells on Earth.

When photosynthetic cells began to evolve they put more and more oxygen into the atmosphere. Other cells had to adapt to the increasing levels of oxygen or go extinct. A new type of cell would evolve, called an aerobe. Aerobes contain enzymes that the cells use specifically for neutralizing the toxic effects of oxygen and free radicals. Some cells never developed this ability. As the atmosphere became more oxygenated these cells retreated to environments lacking oxygen. The anaerobic prokaryotes alive today are descendents of the most ancient of all cells. They live only in oxygenless pockets of soil, deep sediments, still swamps, and inside the digestive tract of animals. Most archaea are anaerobes and represent our ancient cellular ancestors.

The following list gives you the key features that bacteria and archaea share and the ways in which they differ:

➤ Both bacteria and archaea measure about 1-3 micrometers in diameter (round varieties) or 2-4 micrometers in length (rod-shaped varieties).

➤ Both bacteria and archaea can be round (cocci) or rod-shaped (bacilli), depending on species. Bacteria also come in rigid corkscrew shapes (spirila), flexible spirals (spirochetes), or curves (vibrios). Both bacteria and archaea have species that grow in all sorts of irregular shapes; these species are called pleomorphic for "many forms."

➤ Bacteria and archaea can live in aerobic environments (bathed in air) or anaerobic environments (airless).

➤ Bacteria have a cell wall reinforced by the polymer *peptidoglycan*; archaea have a cell wall supported by a different polymer called pseudomurein.

➤ Bacterial membranes contain fatty acids of a straight-chain shape; archaeal membranes contain massive branched-chain fatty acids.

➤ Many archaea can produce methane gas from carbon dioxide; bacteria lack this ability.

➤ Several bacteria can convert the Sun's energy to chemical energy through photosynthesis; archaea lack this ability.

➤ A small percentage of bacteria cause disease in animals and plants; archaea are not known to cause any diseases.

Many of the physical peculiarities of archaea help them thrive in very harsh environments. Archaea have been found in places where no other organisms survive for long. Archaea live in salt lakes, hot sulfur springs, *hydrothermal vents*, acid runoff from ore mines, and inside the digestive tract of animals and insects. These forbidding habitats are called extreme habitats, and any organism that can live there is an *extremophile*. Nonextremophile archaea also inhabit more mundane places like soil and seawater.

Inside Prokaryotes

Few things clutter the interior of a prokaryotic cell. The contents inside the membrane is filled with cytoplasm. All cells also have a region where the densely packed chromosome resides. This region is called the nucleoid. Many ribosomes are also distributed throughout the cytoplasm.

Certain specialized prokaryotes manage to fit in extra structures that support their lifestyles. Thus, some prokaryotes have the following:

➤ Gas vesicles—Sacs that hold air and help maintain the cell's buoyancy in water.

➤ Photosynthetic vesicles—Packets containing materials for carrying out photosynthesis.

➤ Carboxysomes—Sacs that hold carbon dioxide.

➤ Magnetosomes—Tiny magnets arranged in chains that help a cell orient itself to the Earth's poles.

➤ Storage granules—Packets of stored food, such as sugars or glycogen.

Outside Prokaryotes

The tough cell wall lies just outside the fragile plasma membrane. This cell wall takes the hits doled out by the cell's environment so that the membrane will stay intact. Should the membrane break, the cytoplasm will gush out and the cell dies.

Prokaryotes come in two varieties based on the cell wall's composition. In the 1800s, Danish microbiologist Christian Gram invented a way to stain bacteria so that they would be easier to view in a microscope. Gram had also unexpectedly discovered that all bacteria seem to behave one of two ways when exposed to the stain. These two groups became known as gram-positive and gram-negative species. The Gram stain remains a basic method used by microbiologists to identify microbes from the environment or to diagnose an *infectious disease*.

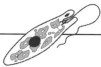

Inside a Cell

Gram-positive and Gram-negative

In a microscope, gram-positive cells are dark blue or purple due to a cell wall that traps the stain crystal violet. Gram-negative cells remained unstained and colorless. To make gram-negative cells as easy to see, the Gram stain method calls for a step in which the cells are exposed to the stain safranin. This turns gram-negative cells pink.

Prokaryotes possess a variety of structures attached to the outside of their cell wall. These structures serve to sense changes in the environment, communicate with other cells, attach to inanimate surfaces, or attach to another living thing just prior to starting an infection.

Microbiologists are discovering new layers and appendages on prokaryotes even now, but the most well-known are:

➤ Capsules—Polysaccharide layers overlaying the cell wall. Capsules retain moisture for the cell during dry conditions and also help the cell avoid being destroyed by the body during an infection.

➤ Slime layers—Polysaccharide layers similar to capsules but more easily washed off in water. Slime layers give cells some level of protection from damage in the environment.

➤ S-Layers—A protective shell-like layer on many archaea, which in a microscope resembles a tile floor.

➤ Mycolic acid—A waxy substance that lies outside the cell wall in certain disease-causing microbes such as the tuberculosis bacterium.

➤ Pili—Short, fine hairs that allow two cells to connect and share genetic material.

➤ Fimbriae—Short hairs that can number as high as 1,000 covering the cell surface, they help cells attach to surfaces.

➤ Flagella—Long, somewhat flexible, tails that enable cells to swim.

Prokaryotic cells often grow in groups. These groups can contain several cells to several dozen. Certain species assemble into a characteristic type of cluster and avoid other arrangements. For example, *Staphylococcus* bacteria always grow in big clumps like a bunch of grapes. They prefer not to form strings of cells lined up end to end. By contrast,

Streptococcus bacteria love to grow into long conga lines. Microbiologists use the following terms to describe the typical ways some species of prokaryotes like to grow:

➤ Diplococci—Two cocci connected as a pair.

➤ Diplobacilli—Two bacilli connected end to end.

➤ Tetrad—Four cocci connected in the same plane.

➤ Sarcinae—Eight cocci connected to form a cube.

➤ Staphylococci—Grapelike clusters or broad sheets of cocci.

➤ Streptococci—Several cocci connected end to end, forming a chain.

➤ Streptobacilli—Several bacilli connected end to end, forming a chain.

Diversity in Prokaryotes

By now you may be realizing that prokaryotes come in a variety of shapes, cell configurations, outer coatings, and appendages. This variety hints at the enormous diversity of prokaryotes.

Not only do prokaryotes have various physical characteristics, but they also carry out a broad range of chemical reactions. The breadth of prokaryotic chemistry is more diverse than the chemistry inside eukaryotic cells. Humans, wolves, alligators, and field mice all share pretty much the same chemistry inside their cells. By comparison, prokaryotes carry out very different chemical activities in a much wider range of habitats.

It may be an almost impossible task to list all the things that prokaryotes do on Earth. They keep all other biota alive. Even disease-causing *pathogens* serve a purpose by keeping populations in check so that an organism does not overrun its available food and space. Here is a list of the general tasks carried out by prokaryotes:

➤ Perform photosynthesis.

➤ Absorb nitrogen from the atmosphere and make it available to plants and animals.

➤ Recycle carbon, nitrogen, sulfur, phosphorus, and metals.

➤ Decompose dead organic matter and sewage.

➤ Produce greenhouse gases that warm the planet's surface.

➤ Digest fibers in animal diets.

➤ Make nutrients in soil available for plants.

➤ Protect the body from infection and strengthen the immune system.

➤ Cause disease in humans, animals, insects, and plants.

> ➤ Serve as food for other microbes, which then feed ever-larger organisms in a *food chain*.

> ➤ Produce useful products such as enzymes, antibiotics, vitamins, and polymers.

> ➤ Corrode metals.

> ➤ Neutralize toxic chemicals in the environment.

Perhaps nothing illustrates the diversity of prokaryotes better than extremophiles.

Extremophiles

Extremophiles are prokaryotes that live in an environment too harsh for most other life. Bacteria and archaea each have species that live in extreme environments, but most extremophiles are archaea.

Biology Tidbits

Hydrothermal Vents: Earth's Harshest Environment

Scientists have known about hydrothermal vents for only about the past 30 years. These vents exist on the ocean floor and spew superhot hydrogen sulfide gas into the surrounding water. The area at the vent opening exceeds 660°F.

At this place lives extreme hyperthermophile prokaryotes as well as red, gutless tube worms that grow to longer than three feet in length. Some of the prokaryotes live inside these worms and produce organic compounds the worms use in their nutrition.

Think of any place that would seem to be torture to a human, and that spot is likely home to an extremophile. No place seems to be too hot or too cold, too acid-filled, or too salty to keep out an extremophile.

Extremophiles specialize in the horrid conditions they prefer. These are the names for the types of extremophiles that have found a home in the Earth's worst habitats:

> ➤ Thermophiles—Live at 104-162°F. (Microbes that live in the comfy temperatures favored by people are called mesophiles.)

> ➤ Hyperthermophiles—Live at 150-250°F.

> ➤ Extreme hyperthermophiles—Live at temperatures hotter than 250°F.

> ➤ Psychrophiles—Live at 17-65°F.

> ➤ Halophiles—Live in salt levels greater than 3 percent.

> ➤ Acidophiles—Live in acidic conditions from mild acids, such as vinegar, to strong acids, such as sulfuric.

> ➤ Alkaliphiles—Live in basic or alkaline conditions such as found in soda lakes.

> ➤ Barophiles—Live under high pressure such as found at ocean depths of 2.5 miles or deeper.

> ➤ Xerophiles—Live in dry conditions such as deserts or the Antarctic.

> ➤ Radio-tolerant—Survive exposure to radiation that would kill a human.

After learning about all of these extreme environments, you might think of the prokaryotes that live in mild conditions as rather dull. The prokaryotes of the same environment where most mammals live are anything but dull. They include all of the pathogens that affect human, animal, and plant health.

Cyanobacteria

Cyanobacteria are a large and diverse group of bacteria that deserve special mention because they perform photosynthesis. Cyanobacteria are thought to be among the most ancient prokaryotes. Their evolution led to the accumulation of oxygen in the Earth's atmosphere. This in turn allowed the evolution of humans and every other oxygen-respiring organism on Earth today.

Cyanobacteria began emerging about 3.5 billion years ago and dominated life on Earth for the next billion years. (Some photosynthetic bacteria preceded the cyanobacteria but they used a type of photosynthesis that does not release oxygen as an end product.) They continue to be some of the most numerous microbes on the planet, living in the surface waters of oceans and freshwaters, on rocks, plants, and in the uppermost layer of soil. Intrepid microbiologists have also found cyanobacteria growing in fountains, mineral springs, salt marshes, and glaciers.

Microbiologists have long been puzzled by cyanobacteria because of their diverse cell forms. Cyanobacteria can be single cells, large aggregates, or long filaments. For a long time, cyanobacteria were called blue-green algae because of their color and because their photosynthesis fooled biologists into thinking they were algae.

Along with algae and green plants, cyanobacteria replenish the air with the oxygen we need to survive. Though difficult to estimate, all the Earth's cyanobacteria may produce more oxygen than the total of green plants and trees.

The Domains of Living Things

Bacteria and archaea emerged early in evolution from the same common ancestor. Their evolutionary paths then split from each other early in the first billion years of life.

Today biologists recognize three *domains* of life that have followed three distinct paths from a common starting point in evolution. These domains have been determined by analyzing the differences and similarities of species. The formal names of the three domains are:

➤ Domain Bacteria—All the prokaryotes, including cyanobacteria, with the exception of archaea.

➤ Domain Archaea—All the prokaryotes with the exception of bacteria.

➤ Domain Eukarya—All the eukaryotes, which encompasses protozoa, algae, fungi, invertebrates, insects, reptiles, amphibians, plants and trees, and animals.

CHAPTER 4

The Eukaryotic Cell

In This Chapter

➤ A description of how the first eukaryotes formed.

➤ The main eukaryotic cell structures, called organelles.

➤ Major types of protists and their features.

➤ How single-celled eukaryotes relate to the first green plants

---

In this chapter you will take a walk through a eukaryotic cell. The easiest way to remember which organisms are composed of eukaryotic cells is to first picture in your mind the bacteria and archaea. Now eliminate them from your thoughts. Everything else is a eukaryote.

Eukaryotic cells are more complex than prokaryotes; they represent a step forward in evolution toward today's higher organisms.

Both plants and animals consist of eukaryotic cells. This chapter will show you the many similarities between the two and the ways in which they differ.

Following custom so far in this book, I'll begin with a brief explanation of how eukaryotes developed on Earth.

The First Eukaryotes

The first precursors to prokaryotic cells appeared 3.5 billion years ago. These cell ancestors were little more than tiny balls formed by fatty compounds called *phospolipids*. Inside the balls existed small amounts of peptides and RNA. Over the next 2.5 billion years, the cells became more sturdy and self-sufficient. The peptides elongated into functioning proteins and the RNA became more complex and led to DNA.

Prokaryotes diversified into photosynthetic types and non-photosynthetic types. The photosynthetic cells could survive in nutrient-rich oceans as long as they received sunlight. Non-photosynthetic cells needed an energy source other than sunlight. Some of these prokaryotes hunted other, smaller cells as food. If they were a type that had evolved a tail, or flagellum, the task of chasing down their food became even easier.

The first eukaryotes likely evolved from prokaryotes that caught and ingested other prokaryotes. Most of the time, a predator cell would digest the smaller cell it had just consumed. But sometimes, pieces of the ingested cell remained intact inside the predator. On occasion, these cell fragments served a purpose in the predator's physiology. Over time, predator cells that had ingested other prokaryotes became more complicated due to the bits and pieces of ingested cells they carried. A new type of cell developed. This new cell contained internal structures that carried out certain functions. In this way, the organelle of eukaryotic cells evolved.

While prokaryotes symbolize nature at its most efficient, eukaryotes have a more complicated way of life. Think of a eukaryote as a temperamental Ferrari Testarossa and a prokaryote as a bicycle. They both get you there, but the car has more things in it that must all work together perfectly to make it purr.

What Is an Organelle?

If eukaryotes began as prokaryotes swallowed up by other prokaryotes, it would explain why eukaryotic cells' internal structures have membranes. Other than those dot-like protein factories called ribosomes, which I'll discuss in Chapter 13, a eukaryote's internal structures always reside within a membrane.

An organelle is any membrane-bound structure of a eukaryotic cell that carries out a specific function. The main organelles of eukaryotes are:

> ➤ Nucleus—Location of the chromosome, it also contains one or more non-membrane-bound areas called nucleoli where the cell manufactures ribosomes.

> ➤ Ribosome—Small units dispersed in the cytoplasm that act as the site of protein synthesis; can number many hundred ribosomes per cell.

➤ Mitochondrion—This is the site of *respiration* and energy generation.

➤ Golgi apparatus—This organelle makes various cellular products, sorts them, and packages them for export to the outside in sacs called Golgi vesicles.

➤ Endoplasmic reticulum (ER)—This network of connected sacs and tubes serves as the site of membrane synthesis and also holds ribosomes. Rough ER is studded with ribosomes; smooth ER lacks ribosomes. Cells contain both types of ER.

➤ Lysosome—The lysosome is an enzyme-filled sac that digests food.

➤ Centrosome—This organelle produces slender tubules that guide the process of cell division.

➤ Peroxisome—This sac produces hydrogen peroxide from excess oxygen molecules and also degrades the hydrogen peroxide to protect the cell from damage.

➤ Chloroplast—This organelle serves as the site of photosynthesis.

➤ Vacuole—A very large sac in some cells that stores excess food, digests some foods, and breaks down wastes.

In the above list or organelles, lysosomes exist only in animal cells and chloroplasts and vacuoles exist only in plant cells. In Chapter 5 you'll learn about the organelles in more detail.

Except for ribosomes, the above list of organelles describes the main features that make eukaryotes different from prokaryotes.

Biology Vocabulary

Flagella

Flagella (singular: flagellum) are long appendages cells use for swimming and direction. These tails connect to the cell using several interlocking proteins. The flagella of prokaryotes and eukaryotes have different proteins and have different internal architecture. They both have the same overall function, however, which is to propel a cell through its environment.

The ability to swim is called cell motility and it represents a big step in evolution. A motile organism can hunt for food, escape danger, and seek a mate. A nonmotile organism lives a more vulnerable life exposed to the whims of the environment around it.

Protists

"Protists" is a general term for singled-celled eukaryotes. This group contains a wonderfully diverse collection of cell shapes, flexibility, methods of swimming, tactics for finding and ingesting food, and reproduction.

Protists are a little like prokaryotes because they make their way through life on their own. They find their own food, sense danger and react to it, and reproduce without the need for another cell. But because protists are eukaryotes, they have more complex structures and often more complex life cycles than prokaryotes.

Inside a Cell

Euglena

The genus *Euglena* is the most famous of the group of protists called euglenids. These cells are oblong, measuring about 20 micrometers long and five micrometers wide, and they have a long flagellum at one end. They possess chloroplasts for photosynthesis and a characteristic eyespot at one end, set off to one side of the cell. The eyespot blocks light from certain directions from striking an underlying detector. When *Euglena* shifts its position, light bypasses the eyespot and hits the detector. This configuration enables *Euglena* to move toward light of just the right intensity for its photosynthesis.

Protists are so diverse, it can be difficult to get a handle on characteristics common to all of them. The following list describes the protists most important in biology because of their roles in nature or in health:

➤ Amoeba—A shapeless cell that creeps through water by amoeboid movement.

➤ *Euglena*—Uses photosynthesis when light is available and by other types of *metabolism* in the dark. It contains a large eyespot filled with pigments that help the cell swim toward light and a large flagellum that propels it. This organism is one of the most-studied microbes in biology classes.

➤ Diplomonads—Contain two equal-sized nuclei. The most famous in the group is *Giardia*, which can contaminate natural untreated waters and cause diarrhea.

➤ *Trichomonas vaginalis*—A sexually transmitted microbe.

➤ Dinoflagellates—Always have a double-cell appearance and two flagella.

Dinoflagellates grow to enormous numbers in marine waters and freshwater, sometimes causing *blooms* or *red tides*.

➤ *Plasmodium*—Lives as a parasite inside mosquitoes and humans. When it infects humans, it causes malaria.

➤ Ciliates—Cells covered all over with tiny hairs called cilia. The most famous is *Paramecium*.

➤ Oomycetes—Similar to fungi but they have flagella, which fungi lack. Oomycetes include water molds and the plant pathogens called white rusts and downy mildews.

➤ Algae—Single-celled forms of algae—and there are many—can also be classified as protists. These include diatoms and certain golden, brown, red, and green algae.

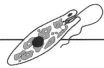

Inside a Cell

Amoeba and Amoeboid Movement

An amoeba (plural: amoebae) is a soft-walled protist of no particular shape. Amoebae move by pushing out a section of their outer membrane and then pouring their cytoplasm into this protuberance. The protuberance is called a pseudopod or "false foot." This slow fluid style of motility is called amoeboid movement.

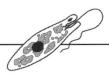

Inside a Cell

Paramecium

The genus *Paramecium* is the most famous of protist ciliates. A *Paramecium* cell is always oblong, constantly takes in water from its environment, and always lives in freshwater. Cells are almost 250 micrometers long and 50 to 75 micrometers wide. Students study paramecia cells because of their distinct organelles and feeding behavior. *Paramecium* eats bacteria by sweeping them using their cilia into an oral groove into a rudimentary mouth.

Algae

Algae are incredibly diverse organisms that range from single-celled protists to massive ocean kelp forests. Within each subgroup of algae exist many thousands of species.

Biologists organize algae into classes based mainly on the pigments the cells contain rather than whether the algae are single-celled or made of many cells. All algae contain at least one type of chlorophyll, the same pigment found in green plants on land. But algae are grouped based on the additional pigments they possess, as follows:

➤ Golden—Photosynthetic algae containing yellow and brown pigments called *carotenoids*, they live in freshwaters and marine waters.

➤ Brown—These brown or olive-colored multicellular (made of many cells) algae are better known as seaweed and live along cool-water coasts. Many seaweeds and kelps have developed an adaptation called a holdfast that anchors them to the shore while being buffeted by tides and waves.

➤ Red—More than 6,000 species of red algae exist. They mostly live in warm, tropical coastal waters and all contain the red pigment phycoerythrin. Some are seaweed. In deep water, species can be almost black. Moving to shallower water, the algae appear red, and in surface waters these organisms are green. A few red algae species have lost their pigments during evolution and must eat other algae for energy.

➤ Green—The green pigment chlorophyll gives these algae their characteristic color. More than 10,000 species exist. Green algae live as single cells, as small balls of cells (such as the organism *Volvox*), and as multicellular organisms of many thousands of cells (such as the organism *Ulva* or sea lettuce).

Biology Tidbits

Volvox

Volvox is a hollow ball formed by the gathering of thousands of green algae cells. In evolution, *Volvox* represents an early attempt by single-celled organisms to join together in a common purpose. *Volvox* cells surround themselves with a gelatinous goo that protects them from damaging substances in their freshwater environment. They also form a rudimentary cell-to-cell communication system by extending strands of cytoplasm. If isolated from its colony, a *Volvox* cell cannot reproduce.

The study of algae could consume a lifetime. Not only are these diverse organisms that form various types of communities, algae also contribute mightily to the atmosphere's oxygen levels through their photosynthesis. Algae and cyanobacteria put enormous amounts of oxygen back into the atmosphere and together, they account for more oxygen than all terrestrial plants and trees.

Single-celled algae and small aggregates of alga cells make up part of phytoplankton. Phytoplankton is the Earth's major site of photosynthesis and contributor of oxygen to the atmosphere.

Diatoms

Diatoms are unique types of algae that perform photosynthesis but also have a hard shell made of silica. Diatoms are always single-celled.

The glass-like shell of a diatom is one of the strongest structures on Earth. Scientists measuring the shell's strength have determined that each shell withstands the pressure under one leg of a table. Did I mention also that the table supports the entire weight of one elephant?

Diatoms offer an added fascination to biologists because of the intricate patterns of the shells. Each shell contains a pattern of lattices, lacework, holes, and grooves. In addition to giving diatoms among the most beautiful appearances in nature, the shapes help increase the strength of each shell. A smooth or flat diatom shell would break under the pressures routinely shouldered by patterned diatoms.

Biology Tidbits

Diatomaceous Earth

When larger organisms eat diatoms, they cannot digest the silica shell. The predator eliminates the shells, which drift to the sea bottom. For many millions of years, diatom shells have built up a layer of chalky fossils on the ocean floor. Companies harvest this material known as diatomaceous earth (DE). DE is abrasive, so it serves as a cleanser for removing corrosion from metal surfaces and in polishes. DE has also been put to use in toothpastes—again, because of its abrasive action—and in water filters for removing small particles.

Water Molds, White Rusts, and Downy Mildews

Oomycetes are similar to fungi and these two types of organisms have been confused. Here are the differences:

➤ Fungi contain a cell wall made of the compound *chitin*; oomycetes have cell walls made of cellulose.

➤ Fungi lack flagella and seldom live in water; oomycetes need flagella for swimming because they live exclusively in water.

Perhaps people confuse oomycetes and fungi because of their similar fuzzy appearance. Oomycetes mimic fungi by producing long filaments of growth that eventually turn into a tangled mass. By producing filaments, these organisms increase their surface-to-volume ratio and thus increase the chance of taking in nutrients from the environment.

Although many of the oomycetes annoy gardeners and commercial growers because they cause plant diseases, oomycetes also act as important decomposers of waste. Any dead plant or animal in water will probably be decomposed by an oomycete.

Protozoa

Protozoa are protists that specifically get food by catching it and ingesting it.

Protozoa do not have the ability to use photosynthesis. Some microbiologists prefer the term "protozoa" rather than "protists" to describe any non-photosynthetic eukaryotic cell. By this definition, protozoa include everything I have discussed in this chapter except the algae.

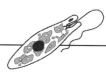

Inside a Cell

Flagellates and Ciliates

People who study protozoa for a living group protozoa based on cell locomotion. Two major ways of swimming are through the use of long tails or flagella or by many short hairs or cilia. Protozoa can therefore belong to the group loosely termed flagellates or the ciliates. Both types swim forward and backward, change direction quickly, and go fast! If you were to sit at a microscope to observe a drop of water filled with flagellates and ciliates, they would whiz by in the blink of an eye!

Protozoa must hunt and snatch their food in their environment. Some protozoa, such as flagellates and ciliates, shoot through their surroundings to catch bacteria and even smaller protozoa. Others, such as amoebae, lumber along at a sloth-like speed. But even amoebae envelop the food they come upon and ingest it.

Many protozoa cause illness in humans and other animal life. The following diseases are caused by a protozoan: dysentery, malaria, babesiosis, toxoplasmosis, leishmaniasis, giardiasis, and cryptosporidiosis.

Algae Lead to the First Green Plants

The first plants arose from two algae that depend on chlorophyll for a big part of their photosynthesis: red algae and green algae.

Red algae developed a complex lifecycle in their evolution. This life cycle is perhaps a precursor to the lifecycles of higher plants. Meanwhile, the pigments and chloroplasts of green algae are very similar to today's green plants. Molecular biologists who study the genetic makeup of organisms have determined that green algae and land plants are closely related.

Biology Vocabulary

Phytoplankton

Phytoplankton is a group of organisms consisting mainly of cyanobacteria and algae. Phytoplankton drifts in the upper surface layer of water environments. From this location, the cells absorb the Sun's rays and turn solar energy into chemical energy that other life on Earth use.

"Phyto-" refers to the plant-like nature of these microbes. By contrast, zooplankton consists of tiny animal life, such as invertebrates that live in water. Zooplankton cannot photosynthesize, but they do serve biology by being a food source for other aquatic life.

Together, phytoplankton and zooplankton make up the diverse group of microscopic and very tiny organisms called plankton.

CHAPTER 5

How Your Cells Work

In This Chapter

➤ The basics of membrane function and types

➤ Main ways in which nutrients go through membranes

➤ The major energy-generating structures of cells

➤ A tour of mitochondria, the nucleus, and other important organelles

In this chapter you will learn how all the cells of your body power you through the day. Imagine strolling through a park in search of a shady spot to relax, or competing in a triathlon, or dashing to the corner deli for a pastrami on rye. How do the 100 trillion cells that make up your body generate the energy for reading this book, cycling 100 miles, or digesting a sandwich? This chapter introduces you to the machinery of the human cell.

By knowing your cells, you will gain an appreciation of how to care for them, feed them, and give them sufficient rest. It all begins with the power-generating organelles of each cell. These are mainly the membranes and the mitochondria. The energy made in these places enable cells to take in more nutrients, grow and reproduce, communicate, and carry out their specialized functions in the body.

The energy generated by a cell thus helps the workings of the other structures of the cell. This chapter will cover the nucleus, Golgi apparatus, and the endoplasmic reticulum, plus a few other organelles that carry out special tasks for you. Now let's take a closer look at your most important cell organelles.

Biological Membranes

All cells depend on membranes for three life-giving functions. First, membranes protect the cell interior from the sometimes harsh realities outside the cell. Second, membranes control what gets into a cell and what goes out. Third, membranes house energy-generating systems. In eukaryotes, membranes also provide a support infrastructure for ribosomes, the cell's protein factories.

Animal and plant cells have four main membrane systems:

> ➤ Plasma membrane—Encloses the cell cytoplasm and also does the following jobs: (a) transports nutrients into the cell; (b) excretes wastes from the cell; (c) releases specific cellular secretions; (d) maintains the cell's ion balance; (e) provides a site for energy-generating steps; and (f) communicates with other cells.

> ➤ Endoplasmic reticulum—The ER offers a site where the cell manufactures new membrane and also provides a place for ribosomes.

> ➤ Nuclear membrane—Sometimes called a nuclear envelope, this is a double membrane that surrounds the cell nucleus. The nuclear membrane is also continuous with the ER. This connection helps send genetic information from the nucleus to the ribosomes, which turn genetic information into functioning proteins.

> ➤ Mitochondrial membrane—Another double membrane, this encloses the mitochondrion, the cell's main energy-generating site.

Biology Vocabulary

Ions and Ion Balance

An ion is an atom that has gained one or more extra electrons or lost one or more electrons. The result of this gain or loss is a charge. For example, the element sodium (Na) has no charge. The sodium ion (Na^+) holds one less electron and therefore has a positive charge.

Many nutrients cross biological membranes as ions rather than in a neutral form. The plasma membrane plays the essential role of maintaining the cell's ion balance. This means the membrane continually shunts ions back and forth so that the cell's interior charge nearly matches the charge on the exterior.

Lipid Bilayer

Biological membranes are called lipid bilayers because they consist mainly of fatty acid lipids arranged in a double-sided configuration similar to a sandwich. The outer surface of each membrane in a cell usually interfaces with an aqueous mixture. For example, blood cells contact blood, which is more than 90 percent water. The cells that make up organs come in contact with blood or cerebral spinal fluid or *lymph*. By contrast, the interior layer of membranes repels water.

Membranes have been described as fluid-like because they are not static structures. The fats and proteins of membrane flow about within the membrane with quite a bit of freedom. Even the largest proteins embedded in membranes can slowly drift from place to place in the membrane.

The fluid nature of membranes makes them permeable. This means that certain substances flow through membranes. Of course, if everything went through our membranes willy-nilly, we would rapidly succumb to drug overdoses, chemicals, excess nutrients, and poisons made by infections. The fluid-like bilayer prevents wholesale movement of substances across the membrane. This ability to grant access to some substances and block other substances from crossing the membrane is called selective permeability.

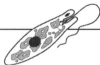

Inside a Cell

Microvilli

Many cells have extensive folding of the outer membrane that creates a landscape of projections and canyons. The projections are called microvilli. Microvilli greatly increase the surface area of a cell. This becomes especially important in cells that are responsible for absorbing vital nutrients, such as the cells that line your digestive tract. The microvilli of cells lining the intestines are so extensive, they create a hair-like surface called the "brush border." The brush border is your body's main location for nutrient absorption.

Every second, your cells shunt several substances back and forth across the plasma membrane in a constant struggle to keep things in balance. Scientists call this state of balance homeostasis. From cells to organs to entire bodily systems, your body's primary goal is to maintain homeostasis.

The main substances that are constantly traversing the membrane are: water, oxygen (in), carbon dioxide (out), and the ions sodium (Na+), potassium (K+), calcium (Ca+), and chloride (Cl-), which is the element chlorine's ion.

How Substances Go Through Membranes

Every substance that comes in contact with our cells either gains entry or is blocked. Nutrients and many drugs enter. Various chemicals, pollutants, and poisons should be blocked, but scientists are learning that quite a few of these substances make it across the membrane too.

Substances cross the membrane by the following four main mechanisms:

> ➤ Diffusion—Small molecules such as water and oxygen simply flow right through the membrane. This chemical process is called osmosis.

> ➤ Facilitated diffusion—Ions and a few other substances can flow through the membrane, but the crossing works best when helped by a protein already in the membrane. The protein forms a pore through which the substance can pass… kind of like the Suez Canal. Amino acids and some sugars use facilitated diffusion.

> ➤ Active transport—Other than being a good name for a moving company, active transport moves substances that would normally not cross the membrane. In active transport, a protein grabs a substance on the cell's outside and pushes it into the inside. This requires energy. To make up a portion of the energy cost of active transport, the cell simultaneously pumps sodium and potassium ions across the membrane. By pumping sodium out of the cell, the membrane creates a difference in the inside-outside sodium concentrations. This difference, called a *gradient*, can be converted into the energy needed to transport things. But remember, the cell also must keep its charge between the inside and outside balanced. So as a final step, the membrane brings potassium ions (K+) inside to balance the loss of the sodium ions (Na+). During rest periods, the cell rebalances its sodium-potassium levels. Cells use active transport for sugars, amino acids, peptides (short chains of amino acids), and some vitamins.

> ➤ Endocytosis—Like an amoeba, cells engulf big particles and swallow them whole. Cells use this process for taking in macromolecules such as many vitamins, lipids, and large carbohydrates.

Energy Generation in Membranes

The cell produces energy by burning the sugar glucose while consuming oxygen. In a word, this is respiration.

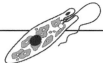

Inside a Cell

Osmosis

Osmosis is the diffusion of water across a selectively permeable membrane. What all these fancy words mean is that when water concentration is low inside a cell, the cell becomes more concentrated with sugar, amino acids, and other stuff. In that case, water flows into the cell. When too much water builds up inside a cell and begins to dilute substances in the cytoplasm, the water flows out of the cell.

Respiration occurs in several connected steps that take place mainly in the membranes of mitochondria and, to a lesser extent, the plasma membrane.

In all organisms, energy production comes from burning mainly carbohydrates and fats. In aerobic organisms—organisms that consume oxygen—respiration acts as the energy-producing system. In anaerobic organisms (bacteria, archaea, or protozoa), *fermentation* is an important energy-producing system.

As cells burn carbohydrates and fats for energy, electrons (e-) and protons (H+) are liberated from the energy reactions. The membrane gathers the electrons and passes them from one specific compound to the next in an electron shuttle system. This shuttle is called the *electron transport chain* (ETC). The ETC can produce energy by creating a difference in electron and proton concentrations inside the cell compared with outside. The inside-outside difference in negatively and positively charged atoms causes a voltage difference across the membrane. This so-called membrane potential is critical not only for generating energy but also for nutrient uptake, cell-to-cell communication, and signal transmission along nerve cells.

Biology Tidbits

Membrane Potential

Membrane potential is the difference in electrical charge from the cell's cytoplasm to the fluid on the outside of the cell. By separating the charges of the cell's inside and outside, the plasma membrane gives the cell a potential source of energy. Think of membrane potential as a battery. Batteries produce energy by sequestering electrons in one location and then allowing them to flow to a second location. The electron flow causes a release of usable energy.

The membrane potential set up by the ETC enables the cell to make a highly energized compound called adenosine triphosphate (ATP). ATP is the main energy currency used by all cells in biology.

How Membranes Help Cells Communicate

Membranes help cells communicate in two ways. The first way is by membrane potentials that move along the cell's outer membrane. You will learn more about this phenomenon in the next chapter in the description of how nerve cells work.

Membranes also help cells recognize each other. Neighboring cells can tell whether their neighbor is of the same type, such as two adjacent skin cells, or of a different type, such as a nerve cell and a muscle cell. This recognition is due to carbohydrates that stick out from the plasma membrane.

Cell-to-cell recognition is vital to your health for the following reasons:

➤ Sorting of cells into tissues and organs during an embryo's development.

➤ Identification of foreign cells during infection so that the immune system can destroy the interloper.

Cell-to-cell recognition processes also cause rejection of transplanted organs. When cells cannot recognize the new group of cells, they alert the immune system. The immune system in turn begins a series of steps that lead to rejection of the transplanted organ.

Biology Vocabulary

Adenosine Triphosphate (ATP) = Energy

ATP is the main substance used by cells to transfer energy to and from various chemicals. ATP carries this energy in the chemical bonds that attach phosphates to *adenosine*. When a phosphate breaks off from the main structure, a small spurt of energy becomes available by which the cell can do work.

Mitochondria

Mitochondria serve as the cell's main site of respiration. Inside mitochondria, the energy contained in carbohydrates and fats is converted to energy in the form of ATP, and oxygen is consumed.

Mitochondria occur in every type of cell except prokaryotes. Most cells have more than one mitochondrion; many cells have hundreds or thousands of them. Cells that must produce movement or do other types of heavy work, such as muscle, contain many mitochondria. Cells that have a more sedentary lifestyle, such as a skin cell, can hold fewer mitochondria.

Each mitochondrion has two membranes. The outer membrane is fairly smooth, but the inner membrane is convoluted and contains many folds. The folds are called cristae. Cristae are the site of the ATP-generating electron transport chain.

The inner membrane encloses a fluid called the mitochondrial matrix. This matrix contains many ribosomes, enzymes, and DNA distinctive to this organelle.

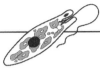

Inside a Cell

Mitochondrial DNA

Every mitochondrion contains a small piece of DNA called mitochondrial DNA (mtDNA). In humans, mtDNA represents a small fraction of a cell's total DNA, but it nevertheless plays a role in human genetics. The genes contained in mtDNA work in making new mitochondria, energy generation, and protein production. Researchers have also linked some mitochondrial genes to hereditary diseases. For example, certain mutations in mtDNA have been implicated in causing retinitis pigmentosa. This is a condition in which cells that detect light begin to degrade prematurely and cause vision loss in childhood or early adulthood. Some mutations of mtDNA are also thought to play a role in the normal aging process.

The Nucleus

Like mitochondria, the cell's nucleus is surrounded by a double membrane. The nucleus contains most of your DNA and therefore most of your genes. The DNA is packaged into discrete units called chromosomes. A typical human cell has 46 chromosomes. Cells involved in reproduction, that is, the egg and the sperm, contain 23 chromosomes each and make a full complement when the female cell (the egg or ovum) fuses with a male cell (the sperm).

The nucleus also contains a dense region called the nucleolus. In the nucleolus, the protein-manufacturing apparatus of the cell gets its instructions from your DNA.

Endoplasmic Reticulum (ER)

As mentioned in the previous chapter, the ER is a extensive membrane system that is folded up inside your cells. Two types of ER exist, each with their own functions:

> ➤ Smooth ER—Functions in lipid production, carbohydrate use, and inactivation of drugs and poisons.

> ➤ Rough ER—So-called because it is studded with ribosomes, this ER acts as the site of protein production. In your body, this includes enzymes, hormones, *globulins*, muscle, and other proteins.

Once proteins are made by the ribosomes attached to the rough ER, the ER breaks off bits to form sacs called transport vesicles. These vesicles carry the newly minted proteins to where they are needed in the cell.

Golgi Apparatus

If a protein or other substance is to be shipped outside the cell, transport vesicles travel to an organelle called the Golgi apparatus, which is responsible for packing up the shipment. Your cells have their own Shipping Department!

The Golgi apparatus consists of stacks of flattened membranes that have been described as a stack of pita bread. Cells can contain hundreds off these stacks.

The organelle frequently breaks of a piece of its membrane to form a shipping vesicle, which then moves to the outer plasma membrane and fuses with it. The cellular products inside the vesicle either join the membrane or continue to the cell's exterior. When they are shipped outside the cell, these products are called secretions.

Other Parts of Your Cells

A battery of lesser-known organelles play individual parts in cell physiology. Some are mainly for structural support, such as the components of the cytoskeleton. Other organelles have tasks related to maintaining the cell's health, such as peroxisomes and lysosomes.

Peroxisomes

Peroxisomes are small round sacs that hold enzymes that protect the cell from damage from oxygen. The peroxisome enzymes put hydrogen atoms on any oxygen atoms released by cellular reactions. This action produces hydrogen peroxide (H_2O_2). Because hydrogen peroxide is also

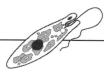

Inside a Cell

Cytoskeleton

The cytoskeleton is a network of fine tubes and filaments that give the cell its structure and maintains its shape. In some motile cells, the cytoskeleton also helps the cell move from place to place.

dangerous to the cell, a second enzyme quickly converts it to water. The overall benefit of the peroxisome is therefore to protect cells from their own by-products of respiration.

Lysosomes

Lysosomes also protect the cell because of the digestive enzymes contained inside these sacs. These enzymes degrade large molecules that get inside the cell and may be damaging to the cell. Certain protists, such as amoeba, rely on lysosomes to digest the substances they take in as food.

CHAPTER 6

How Cells Communicate

In This Chapter

➤ Basic components of cell-to-cell communication

➤ The connection between cell communication and multicellular organisms

➤ An introduction to nerve cells

➤ The role of genes in formation of multicellular organisms

In this chapter you will see how cell communication developed in biology. The first cell-to-cell communications among prokaryotes led the way to more sophisticated communications among cells of complex organisms.

Cell communication is critical for a living organism because it helps neighboring cells coordinate their activities. In prokaryotes, this coordination of activities occurs even while the cells live independently from each other. Cell communication allows many species of prokaryotes to coordinate their actions so that a population of independent cells look a lot like a single organism.

In complex multicellular organisms, cell communication and coordination produces the images you know so well. Imagine a cheetah streaking over the African savannah or a skier schussing down a mountain. These actions are the culmination of trillions of cells working together. They send and receive signals, make instantaneous adjustments, and churn out the energy needed to keep the organism going.

Cell communication also works in ways unrelated to physical exertion. Communication enables an organism to sense conditions in the environment. The sensations are transmitted to the brain, and signals are passed along a communication network that allows the organism to respond to its surroundings. Hundreds of thousands of these signals shoot around your body without your knowledge. The cell-to-cell communication of your body runs your metabolism and determines your behavior. Simply, it makes you who you are.

The first signals passed back and forth between ancient prokaryotes started life on a road that led to humans and every other multicellular being.

This chapter begins with the communication systems of prokaryotes. The chapter then introduces you to cell communication of complex organisms. It all begins with the earliest tentative attempts by cells to sense the world around them.

Chemotaxis

Chemotaxis is a process of prokaryotes for detecting and moving toward desirable chemicals (attractants) and moving away from undesirable chemicals (repellants). Imagine what you would do if you smelled the scents coming from a bakery and compare that to your response to the odors from a sewage treatment plant. Which would attract you and which would repel you?

Biology Vocabulary

Feedback

Feedback is the process of detecting a condition outside a cell and sending this information to the cell's inner workings so that it can respond appropriately. An important component of feedback in biology is feedback inhibition. In inhibition, a cell's end products build up and the cell detects this buildup using its chemoreceptors. The cell then adjusts its metabolism to slow down the formation of the end product. Feedback inhibition thus keeps a cell from being poisoned by its own wastes.

Although chemotaxis is used by the simplest cells in biology, it is far from a primitive system. Chemotaxis in bacteria is one of nature's most complex systems. It involves several proteins that receive signals from the environment, other proteins that interpret the signal, and yet other molecules that temper the cell's reaction to the outside world. By using chemotaxis, bacteria swim with a purpose rather than drift aimlessly.

Let's look at chemotaxis in a bit more detail. On the outside of every bacterial cell sits an array of proteins that detect certain chemicals. These proteins are called *chemoreceptors*. When a chemoreceptor binds with a specific chemical in the cell's surroundings, it starts a cascade of signals inside the cell that tell the cell to swim forward or retreat. This internal response to conditions outside the cell is called feedback. Under a microscope, a bacterial cell seems to tumble aimlessly for a while, then swims in a straight line, and then tumbles again. Although much of the swimming is random, the straight-line motion eventually adds up to moving the cell in a definite direction.

Quorum Sensing

Prokaryotes use chemoreceptors for a specialization called quorum sensing. In quorum sensing, cells monitor the density of their population by detecting chemicals emitted by their immediate neighbors. A high concentration of chemicals—lots of chemoreceptors activated at the same time—means the cell density is high. A low concentration means the population is sparse. If you used this system and lived in New York City, your chemoreceptors would be hyperactive. By comparison, if you lived in South Dakota, your chemoreceptors would not be busy.

Quorum sensing marks an important step in evolution because it shows how independent cells began to live as a community, respond to each other, and communicate with one another.

When bacterial populations reach a critical level of cell density, a portion of the population either dies or swims away to a new location. In this way, the communication links set up by quorum sensing protect and sustain an entire population. By protecting populations, the bacterial species survives.

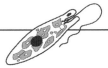

Inside a Cell

Chemoreceptors

The chemoreceptors of prokaryotes detect nutrients as well as harmful chemicals, and they work in similar fashion to the receptors on your cells. In animals, chemoreceptors monitor water levels in the blood (part of what makes you thirsty), oxygen, carbon dioxide, glucose, and amino acids. Many insects and animals also have chemoreceptors that detect sex hormones emitted by others of their species.

Chemoreceptors are a type of sensory receptor in the body. Other sensory receptors exist for touch, pressure, and stretching, electromagnetic energy such as light and electricity, temperature, and pain.

Multicellular Organisms

Quorum sensing represents an early step in the evolution of many-celled organisms. In prokaryotes, quorum sensing occurs right within a bacterial *colony*. Cells at different locations in a colony that measures no more than an eighth of an inch across can behave quite differently from each other. Perhaps the cells along the colony's extremity have the job of gathering in nutrients. Cells in the middle of the colony put more emphasis on storing nutrients. The outer cells crank up their absorption mechanisms to the max while the mid-colony cells spend their time building starch. By taking on different jobs within the same colony, bacteria help the entire population work at its best.

A bacterial colony seems to have little else in common with a complex organism like a human made of specialized tissues, organs, and systems. But the differentiation that goes on inside a speck of a bacterial colony tells us something important about biology: cells differentiate. When they do, they greatly open the possibilities of what nature can produce. You can sum it up in the single word "diversity."

An Early Multicellular Organism?

Myxobacteria show us a possible bridge in evolution between prokaryotes and complex organisms made of millions of cells. These bacteria glide through soil in search of food. When their nutrient supply runs low, they stop moving around. Chemoreceptors relay to the myxobacteria that the chances of starving have greatly increased. The myxobacteria waste no time in responding to this dire message. They start to pile up on top of each other and form a tall structure called a fruiting body.

Fruiting bodies resemble red, yellow, or brown fungi. They look a little like miniature mushrooms. If nutrients completely disappear, a fruiting body makes tough, resistant myxospores. Myxobacteria spores can remain dormant for up to 10 years, waiting for favorable conditions to return. When conditions improve and nutrients again begin to flow toward the fruiting body, it releases the myxospores into the environment. They then wake up and start a new generation of actively growing cells. Therefore, myxobacteria look like regular bacteria when conditions are good, but they turn into an entirely different-looking multicellular creature when conditions turn menacing.

Nature has produced few other species that behave like movie stars (see *Transformers*!). Yet nature's ability to differentiate cells of the same species like those of myxobacteria, gives us clues to the way multicellular organisms evolved.

The oldest known fossils of multicellular organisms made of eukaryotic cells date to about 1.2 billion years ago. These first multicellular organisms were likely colonies of cells, similar to the colonies formed by prokaryotes. In the colonies, different cells became specialized for various tasks. For example, earliest algae produced massive colonies containing cells doing various jobs. Some cells concentrated on photosynthesis; others had the job of absorbing specific nutrients from the environment; others spent their time sensing the environment's physical features. This differentiation was a precursor to further differentiation we see today in animal organs and the components of higher plants.

Biology Vocabulary

Cell Differentiation

Cell differentiation is the development of cells of the same species into different types with various structures and functions. The earliest differentiations in evolution led eventually to multicellular algae, fungi, animals, and plants. Differentiation cannot happen without cell division. In turn, multicellular organisms cannot develop based on cell division alone; division and differentiation must both be present.

When a modern animal or plant embryo grows, each cell becomes increasingly specialized in structure and function, a step called commitment. The specialized cells team up with others of the same type. Many same-type cells form a cooperative arrangement that produces a body organ. Many primitive organisms contain different tissue types, as you will see in later chapters. The ability to form organs is, however, unique to higher forms of life.

Cell Signaling

Once cells learn to live together as a group, and then differentiate into specialized tasks, it helps if they can communicate with each other. Communication between cells of the same type helps create fully functioning body organs. Communication between cells of different types helps a complex organism function. In the big picture, cell communication allows an organism to sense its environment, respond to danger, find food, digest the food, find a mate, and reproduce.

All of these complicated actions come down to a basic process called cell signaling. In animal cells, two sources of information tell an embryo's cell when to differentiate and how:

➤ Substances and organelles in the cytoplasm called cytoplasmic determinants. The set of particular determinants that new daughter cells receive during cell division influence the type of cell they will become.

➤ Signals from neighboring cells. Chemicals from other cells influence which genes will be active during cell differentiation and which genes will be inactive.

Cell signaling leads to a process called induction inside the cell. Induction is the ability of one group of embryonic cells to affect the development of another.

Cell signaling occurs in plant cells, too. Most plant cells begin differentiating inside the seed to make a fully formed embryo. As a plant matures, it undergoes additional differentiation to make new organs, such as leaves, petals, and flowers.

Plants receive additional signals from their environment. For example, day length and temperature signal a plant to differentiate certain cells to make flowers.

Neurons

Nerve cells, or neurons, are the basic unit of an animal's nervous system. A neuron consists of a main cell body and two or more long extensions called dendrites or axons. A single dendrite can be more than a few feet in some animals.

Neurons represent the most sophisticated of cell-to-cell communication systems in the body. Neurons communicate with each other to transmit information from the body's extremities to the central nervous system (CNS), consisting of the brain and the spinal column. They also communicate with other tissue types to register pain and other sensations and to give instructions to cells such as muscle.

Inside a Cell

Neurotransmitters

Neurotransmitters are chemicals released by the tip of a neuron's dendrite. The gap between two adjacent neurons is called the synapse. Neurotransmitters travel across this small divide. When a receiving neuron absorbs a neurotransmitter from a signaling neuron, a message within the nervous system can progress. In this way, neurons send information to the central nervous system as well as send information from the brain to other parts of the body.

Since neurotransmitters are vital parts of normal body function and disease, research in this area of biology is intense. Neurotransmitters of which you may have heard are: norepinephrine, dopamine, serotonin, gamma amino butyric acid (GAMA), glycine, glutamate, aspartate, and other neurotransmitters made of peptides.

Neurons must first grow to a length that allows them to reach the tissues with which they communicate. All the details of how dendrites elongate toward their target tissue remain unknown, but researchers know that the growing neuron "reads" signals given off by other cells to determine their route. Once a neuron has grown to its full length, it sends information back and forth in the nervous system using chemicals called neurotransmitters.

Controlling Gene Expression

Cell communication, differentiation, and signaling all connect with how an animal or plant cell uses its genes. The genes that are contained in DNA can be activated, deactivated, or ignored at any given time. In later chapters, you will see how the information carried in genes is converted into proteins that do work in the body. For now, consider these basic laws of *gene expression*, which is the conversion of information contained in genes into a functioning protein:

> ➤ The DNA inherited by an organism contains information that affects the type and amounts of proteins made by its cells.

> ➤ These proteins control all the body's functions.

> ➤ Gene expression includes two phases: transcription and translation. Transcription is the conversion of information in DNA to an intermediary, namely RNA. Translation is the conversion of information from RNA to newly synthesized protein.

> ➤ In embryos, as cells differentiate, they make different proteins that determine a new cell's structure and function.

> ➤ Even though differentiated cells can be unalike in structure and function, they all contain the same DNA.

> ➤ Since they all contain the same DNA, differentiated cells must develop differently based on specific gene expression.

The information in genes is expressed (turned into protein) or not based on the signals you have learned about in this chapter. These signals come from other cells in a developing embryo, which will soon turn into a fully developed multicellular organism. By using chemoreceptors, signals, and feedback, an embryo follows a specific blueprint contained in its DNA. The result will be an adult, carrying its own unique traits and yet also carrying all the information of its parents and ancestors.

Energy

CHAPTER 7

Introduction to Metabolism

In This Chapter

➤ How enzymes keep metabolism going

➤ The main ways in which cells regulate enzyme activity

➤ Energy and work in your body's cells

➤ The role of ATP in energy generation

In this chapter you will gain a better understanding of what the word "metabolism" means. Metabolism covers all the ways in which an organism acquires and uses energy. This is a fairly broad concept and might still be hard to grasp without breaking metabolism down into its component parts.

This chapter presents the important aspects of metabolism and the two main routes of metabolism that move nutrients and energy in opposite directions: anabolism and catabolism. Picture a pile of bricks on the ground. Anabolism is similar to your body building a wall by placing brick after brick in ascending rows. Catabolism is analogous to disassembling the wall brick by brick.

Enzymes

You have heard enzymes mentioned in almost every chapter up to this point. They must be important! Enzymes are proteins. What makes enzymes distinct from other proteins that make up your hair, skin, or muscle is their ability to run chemical reactions in the body.

Two things make enzymes able to run chemical reactions: (1) their exact sequence of amino acids and (2) the way the amino acid chain folds up.

Any enzyme-mediated reaction in the body looks like this:

Substance A + Enzyme → Substance A-Enzyme complex → Substance B + Enzyme

In other words, after helping Substance A turn into Substance B, the enzyme again becomes available to do the same step over and over. Substance A is the starting molecule called the enzyme's substrate. Substance B is called the enzyme's product. This A-to-B route helped along by an enzyme is a *metabolic pathway*.

Important enzymes (most enzymes end in "-ase") in biology are:

> ➤ Amylase—Degrades starch to the sugar glucose.

> ➤ Lipase—Digests triglycerides to glycerol and fatty acids.

> ➤ Protease—Degrades proteins to amino acids.

> ➤ Sucrase—Degrades sucrose to glucose and fructose.

> ➤ Maltase—Degrades maltose to two glucoses.

> ➤ DNAse—Degrades DNA to nucleotides.

> ➤ Phosphorylase—Adds a phosphate to a variety of chemicals.

How Enzymes Run Chemical Reactions

How do the enzymatic reactions in biology differ from chemical reactions run in laboratories or in huge factories? Chemists often get reactions to go by applying heat to chemicals. The heat adds energy to the system and helps chemicals react with each other to form new chemicals.

In biology, living organisms cannot withstand the extreme heat needed to run many of the reactions that sustain life. Enzymes solve this dilemma by making reactions possible without the need for heating up the substrates. They do this by lowering the initial energy needed for transforming Substance A into Substance B. This initial energy input needed to get a biological reaction going is called activation energy. Without your realizing it, your cells are constantly running reactions and at the same time adhering to the laws of thermodynamics!

Enzymes are catalysts in biological reactions. This means that they help a reaction proceed without being used up in the reaction.

Biology Vocabulary

The Laws of Thermodynamics

Thermodynamics is the study of energy changes that occur in all matter on Earth. In biology, living organisms follow two laws of thermodynamics. The first law of thermodynamics states that the energy of the universe is constant. Therefore, energy can be transferred (from chemical to chemical, for instance) or transformed (for example, from light energy to chemical energy), but energy cannot be created or destroyed. The second law states that with each energy transfer or transformation, the universe becomes a little less organized because some energy escapes with each reaction. Usually the energy escapes as heat.

The level of disorganization in the universe is called entropy. Overall, the entropy of the universe is increasing. One glance at your office desk or in your closet will probably confirm this law of thermodynamics!

Coenzymes

Some enzymes require a nonprotein helper to link up with before they can catalyze a reaction. A nonprotein organic molecule that is required by an enzyme to help it run a reaction is called a coenzyme. Most vitamins act in the body as coenzymes.

When the helper needed by an enzyme is an inorganic molecule (contains no carbon), the helper is often called a cofactor. Minerals such as magnesium, iron, and zinc are examples of cofactors you use in your body.

Regulating Enzymes

How does an enzyme know when to turn on so you can burn more glucose or break down a protein? How does it know when to stop so that you don't degrade all your proteins? Several different systems in the body regulate your enzymes so that they catalyze reactions at just the right rate, giving you energy when you need it and storing energy when it is not needed right away.

Unless you hanker for a degree in biochemistry, you need not know the details of enzyme regulation occurring in your cells at this moment. In general, enzymes are either activated or inhibited based on what binds to the active section of the enzyme molecule. This active section is called the enzyme's *active site*.

Biology Tidbits

Work

What is work? When your alarm rings on Monday morning, you have a pretty good idea of what work is. The term "work" is actually related to physics and the laws of thermodynamics. Work is what happens when a force covers a given distance. For example, pushing a car 10 feet is work. Swimming a pool length is work. Lifting a can of beer, or umm… a glass of milk, is work.

A cell does three types of work: mechanical, transport, and chemical. Mechanical work involves movement, such as the beating of a sperm's flagellum. Transport work involves moving substances across membranes; nutrients come in and wastes go out. Chemical work is the running of reactions that require an energy input to go. These energy-demanding reactions are called endergonic reactions.

For the more biochemically inclined readers, here is an overview of the main ways the body regulates enzyme action:

➤ Competitive inhibition—A substance mimics the enzyme's substrate and binds to the same active site used by the substrate.

➤ Noncompetitive inhibition—A substance ignores the active site, but binds to another portion of the enzyme in a way that affects the enzyme's ability to work.

➤ Allosteric regulation—The enzyme's activity at one portion of its structure is affected by the binding of a substance to another portion of the enzyme structure; can be used to either activate or inhibit an enzyme.

➤ Feedback inhibition—Buildup of the enzyme's product causes the enzyme to stop working until the product's levels decrease.

Local conditions inside your body also affect enzymes. Like every other biological thing, an enzyme has an optimal range of conditions where it works best. Outside this range, the enzyme slows down or may stop working entirely. The two main factors affecting enzymes are temperature and acidity:

➤ Temperature—An enzyme's activity increases as temperature approaches the optimal temperature for doing work. Most human enzymes have an optimal range of about 95°F to 104°F. For other organisms, such as archaea that live in hot springs reaching more than 160°F, their enzymes have a much different optimal range than those of humans.

➤ pH—This chemical term refers to the amount of acid in a material. A pH of 7 is neutral—neither acidic nor basic—and most enzymes have a optimal range of pH 6 to pH 8. In humans, a startling exception belongs to the enzyme pepsin. This enzyme digests food in the stomach where acidic gastric juices lower the pH to around 2. Not surprisingly, pepsin's optimal range is also about pH 1.5 to pH 2.3.

Anabolism and Catabolism

As you go through your waking day and while asleep at night, your body's metabolism upshifts, downshifts, and idles. After a meal and when resting, anabolism predominates. In anabolism, small molecules absorbed from the digestive tract and circulating in the blood are assembled into larger storage molecules, such as the following:

➤ Sugars go into glycogen stored mainly in the liver and also in the kidneys to a lesser extent. (In plants, sugars go into starch.)

➤ Fats and excess carbon from carbohydrates go into the body's fat deposits as triglycerides.

➤ Amino acids go into protein.

Specific enzymes help build these large molecules from smaller molecules.

Biology Vocabulary

Energy and Types of Energy

You take in nutrients by eating to give your body a good supply of chemical energy. When it's time to perform work, the energy stored in chemical form is transformed to another type of energy, the energy of motion. This is kinetic energy. Moving objects perform work by imparting motion to other objects. When an object is still, it has no kinetic energy at all, but it instead contains potential energy.

Think of a basketball perched atop a hill. The still ball contains 100 percent potential energy and zero kinetic energy. When pushed toward the precipice, the ball rolls forward and kinetic energy increases while potential energy decreases. As the ball hurtles down the hill, it carries lots of kinetic energy, but its kinetic energy again decreases as it rolls to a stop at the bottom of the hill, and its potential energy also amounts to much less than if the ball were at the top of the hill. Where did the excess energy go? It was transformed to heat energy.

Animals use mainly heat (thermal energy) and chemical energy in metabolism and in work. Plants have the added ability to use light energy. In photosynthesis, the plant converts the Sun's light energy to chemical energy, such as the energy stored in sugars or starch.

When the body needs to do work, it must break down the chemical energy it has stored in glycogen and fats. Catabolism begins. A new set of enzymes breaks down the stored molecules and feeds the smaller molecules into energy-generating pathways. (A pathway is any reaction that involves several enzymes working in sequence. The product of one reaction becomes the substrate for the next reaction.) In animals, catabolism looks like this:

> ➤ Glycogen is broken down to produce the sugar glucose, which is used in a pathway called *glycolysis*. Glycolysis produces a little bit of energy, and its end products then enter a second pathway called the Krebs cycle, which produces more energy. (Some microbes use fermentation in place of the Krebs cycle to make more energy.)

> ➤ Fats are broken down to small carbon compounds that go into the Krebs cycle for making energy.

> ➤ Proteins are broken down to amino acids, which also produce energy by being chewed up in the Krebs cycle.

Anabolism usually requires energy to build large storage molecules from smaller molecules. This energy is put in as heat. A heat-absorbing reaction in biology is called an endergonic reaction. Catabolism, by contrast, releases energy so that you can use this energy for work. Energy-releasing reactions are called exergonic reactions.

Biochemists use the term "free energy" for the portion of energy that can perform work when temperature and pressure are held constant.

Equilibrium

When any system reaches equilibrium, it can no longer do work. Think of a dead car battery as a system in equilibrium. In a dead battery, all of the energy flow inside the battery has become equally distributed between the battery's two charged poles. Put another way, equilibrium is a maximum state of stability. (In biology, the closest thing known to approach this condition is a fellow watching football while sprawled on a couch.)

Your body's cells seldom reach equilibrium, thank goodness. A constant flow of substances into and out of cells keeps a minimum amount of metabolic pathways flowing. Some soak up free energy and others release free energy. This in turn prevents the cell from reaching equilibrium.

ATP - The Cell's Energy Currency

We may think of sugar as a high-energy food, but the sugar molecule does not give us our energy directly. The sugar glucose must first be metabolized by the two pathways mentioned earlier: glycolysis and the Krebs cycle. These pathways lead to the transfer of energy held in glucose to another compound. This other compound occurs in all organisms on Earth as a ready source of energy: adenosine triphosphate or ATP.

In the biological world, ATP is a small molecule. It contains only one sugar (ribose), one nitrogen base (adenine), and three phosphate groups. The cell uses the energy held in the bonds that connect the phosphate groups to each other to power endergonic reactions.

How ATP Stores Energy

The bond that holds the third phosphate group to the adjoining second phosphate group in ATP contains about 7.3 kilocalories (kcal) of energy. This is about the same amount of energy as contained in tablespoon of alcohol or a bite of watermelon.

Because the bonds of ATP release energy usable for the cell, they are sometimes called *high-energy phosphate bonds*. These bonds occur in other places in metabolism where the cell has a need for large amounts of energy.

ATP releases energy when the enzyme ATPase cleaves off a phosphate group. The reaction is called hydrolysis because a water molecule splits at the same time; "hydro-" means water and "-lysis" means that something is being broken apart.

How ATP Performs Work

When a phosphate group comes off ATP, enzymes in the cell transfer it to a variety of other chemicals. When a chemical gets this new phosphate, it is said to be phosphorylated. The chemical now holds a high-energy phosphate bond.

When ATP gives a phosphate group to another chemical, the phosphorylated version of the chemical becomes slightly unstable. By this I mean that the chemical is "unhappy" in its new version and wants to change back to something more comfortable. It's a bit like a male rugby player being asked to don a ball gown. The phosphorylated chemical gives away the extra phosphate group as quickly as possible. In this process, it also releases a spurt of energy. The cell uses this energy spurt for doing work.

How ATP Regenerates

The cells of your body are at work almost continuously, but you don't run out of ATP. You make more ATP with an enzyme that gathers up the phosphate groups coming from the phosphorylated chemicals. The enzyme ATP synthase puts the phosphate back on a chemical called adenosine diphosphate (ADP). ADP is simply ATP that is missing one of its phosphates.

When ATP goes to ADP, energy is released. When it is time for you to regenerate ATP from ADP, you must put energy in. The energy that reconstitutes ATP comes from respiration.

Thus, the ATP-ADP conversion has been compared to a turnstile in which ATP is constantly made, used, and remade in your cells. To keep it all going, you must supply your cells with nutrients (phosphorus is one of them) and absorb oxygen. Nutrients and oxygen make respiration go.

CHAPTER 8

Respiration

In This Chapter

➤ The three stages of respiration: glycolysis, Krebs cycle, and the electron transport chain

➤ The basics of ATP energy

➤ ATP output from the total respiratory process

➤ The role of oxygen in our body

In this chapter you will learn about respiration. This is a vital metabolic pathway because it generates energy for every aspect of human metabolism. In fact, all other organisms that breath oxygen also use respiration. This includes aerobic prokaryotes, protozoa, fungi, invertebrates, insects, fish, and mammals.

The main energy-generating steps in respiration occur in the cells' mitochondria. Some of the preliminary steps, such as glycolysis, occur in the cytoplasm and the end products of glycolysis must then be transported into the mitochondria.

How We Get Energy from Organic Fuel

Your body runs on carbon compounds like a conventional car runs on gasoline. A car consumes oxygen during combustion. In the process, it produces kinetic energy and gives off excess energy as heat. Your body similarly respires oxygen to produce energy and also loses a little of food's energy as heat. That's why eating a meal often warms you up. Both cars and humans give off carbon dioxide as an end product. In respiration, the reaction is:

Organic compounds + Oxygen → Carbon dioxide + Water + Energy

Most of the energy generation from food comes down to very basic chemical reactions called redox reactions. The word "redox" is an abbreviation for reduction and oxidation. These are terms used in chemistry for reactions in which two different compounds exchange electrons with a simultaneous production of energy.

Biology Vocabulary

Redox Reactions

Electron transfers between two chemicals are called oxidation-reduction reactions, or redox reactions for short. (I guess "oxred reactions" didn't sound as good.) Every redox reaction consists of oxidation of one chemical coupled with the reduction of a second chemical. Oxidation equals the loss of one or more electrons from a molecule; reduction equals the gain of one or more electrons by a molecule. Why would gaining electrons be called "reduction"? This is because electrons are negatively charged, so adding one makes a molecule's charge go down. The chemical that offers up its electrons is called a reducing agent and the chemical that accepts electrons is called an oxidizing agent.

Respiration occurs in stages, beginning with glycolysis. Almost every cell on Earth begins its energy metabolism with glycolysis. You therefore begin your energy metabolism in exactly the same way as a bacterial cell buried deep in sulfur-emitting swamp mud!

After glycolysis, aerobic organisms and anaerobic organisms differ in their main routes for generating energy. Let's look at how humans do it.

Glycolysis

Because glycolysis shows up throughout the biological world, biologists refer to it as the "universal pathway." The word "glycolysis" means "splitting of sugar." The sugar here is glucose. This six-carbon sugar is essential to all cells.

Glycolysis consists of ten enzymatic steps that convert one glucose molecule into two molecules of the three-carbon compound called pyruvate. This process produces a small amount of ATP for the cell's energy needs, but its real value is in making pyruvate, which can then be further metabolized in respiration for much more ATP energy. The main pathway for making ATP in aerobes is the Krebs cycle.

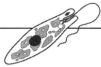

Inside a Cell

Glucose

Glucose is the principal carbohydrate we burn for energy. It also serves as a source of carbon for building new cells during growth, in injury repair, and in the normal replacement of aged cells in the body. Glucose's chemical formula is $C_6H_{12}O_6$.

Glucose serves as the simplest unit in long strings of glucoses that make up either starch or glycogen. Starch is a major energy storage compound in plants and many microbes. Glycogen is a major energy storage compound in animal cells.

When the body needs more glucose than the diet provides, it can reverse glycolysis. By running glycolysis in the opposite direction, the body makes glucose instead of breaking it apart. This reverse-glycolysis route is called gluconeogenesis.

Krebs Cycle = Citric Acid Cycle = Tricarboxylic Acid Cycle

If impressing your friends with your knowledge of biochemistry, you may wish to speak about the Krebs cycle by its other name, the citric acid cycle. Go for even more syllables by calling it the tricarboxylic acid cycle. They all mean the same thing. It is named for the German biochemist and Nobel Prize recipient Hans Adolf Krebs, who identified the cycle's steps in the 1930s.

The Krebs cycle involves eight enzyme-driven steps that take two carbons from pyruvate (from glycolysis) to form a new compound called acetyl coenzyme A, acetyl CoA for short. This pathway is called a cycle because some of the carbons can be reused at the final step when a new acetyl CoA feeds into it.

Fats and amino acids also feed into specific steps along the Krebs cycle. For example, when the body burns fat for energy, it breaks off pieces from fat two carbons at a time as acetyl CoA. This fat-derived acetyl CoA goes into the Krebs cycle exactly the same way as the carbons from glucose enter after going through glycolysis.

Various amino acids from protein breakdown also can be rearranged by enzymes and turned into intermediate compounds of the Krebs cycle. In this way, the body can use protein for energy just as it uses carbohydrates and fats. We use protein for energy only in dire circumstances.

Cells consume water and produce carbon dioxide in many steps of glycolysis and the Krebs cycle. For example, that extra carbon kicked out by pyruvate as it leaves glycolysis and enters the Krebs cycle is exhaled by you as carbon dioxide. These metabolic pathways also produce excess electrons. The electrons go to yet another energy-generating pathway called the electron transport chain.

Biology Tidbits

Carbon Dioxide

Carbon dioxide levels in the atmosphere have risen dramatically since the beginning of the Industrial Revolution. This period coincided with increasing use of combustion engines, which emit carbon dioxide as an end product of burning fossil fuels (gasoline, natural gas, and coal). Respiration in animals and aerobic microbes also contribute to the atmosphere's carbon dioxide. Over geologic time, volcanoes are also a source of carbon dioxide.

Photosynthetic plants, algae, and cyanobacteria remove substantial amounts of carbon dioxide from the atmosphere to slow its buildup. These organisms use a metabolic pathway similar to the Krebs cycle in reverse, called the Calvin cycle. The Calvin cycle takes in carbon dioxide molecules and builds them into glucose. The organism stores excess glucose by stringing together many glucose molecules to make the polysaccharide starch. These processes are anabolic pathways so they require energy input in the form of ATP.

Biology Vocabulary

Cytochrome

A cytochrome is a protein that contains a structure called heme. Heme is a chemical that forms a claw-like grasp around an iron atom. In the electron transport phase of respiration, iron plays a key role in accepting electrons from some compounds and donating them to other compounds. (Heme is also part of the blood protein hemoglobin, except that hemoglobin's iron carries oxygen instead of electrons.)

Electron Transport

The electron transport chain (ETC) is the meat and potatoes of energy production in respiring organisms. Using the ETC, cells harvest every last bit of energy they can from the foods you eat.

Unlike glycolysis and the Krebs cycle, which are based on organic compounds, the ETC is based on proteins. The ETC consists of nine molecules located in the inner membrane of mitochondria. Most of these proteins are cytochromes. Each member of the ETC takes electrons from the member ahead of it and hands those same electrons off to the next molecule in the chain.

Protons (H+) tend to follow electrons around in these reactions of the body. The negatively charged electrons have a natural attraction to positively charged protons. As protons accumulate, the body combines them with oxygen to make water. For this reason, each running of the ETC produces a small amount of water at the same time it consumes oxygen. By using oxygen to take away the excess electrons, the body can continue cranking chemicals through glycolysis, the Krebs cycle, and the ETC.

The ETC also produces ATP using an elegant enzyme-driven system called chemiosmosis. British biochemist Peter Mitchell proposed the theory of chemiosmosis in 1960 as the major way respiring organisms get energy from carbon while ridding their body of excess electrons. (If you want to brag about your electric personality, you now have a biochemical reason to back up your claim!)

Phosphorylation - Big Word with a Big Meaning

In glycolysis and the Krebs cycle, the phosphates that eventually end up on ATP get transferred around from compound to compound. In these two pathways, this process of putting phosphate on an organic compound is called substrate-level phosphorylation. The ETC uses a slightly different version called oxidative phosphorylation. Either way, phosphorylation is a critical component of energy use inside your body. This partly explains why phosphorus is an essential nutrient.

Biology Vocabulary

Chemiosmosis

Chemiosmosis is the part of electron transport that makes new ATP molecules from ADP. As the ETC runs, it pushes protons across the mitochondrion's inner membrane. Protons begin building up on one side of the membrane and the result is a proton gradient.

The gradient releases energy as the protons start to flow back across the membrane to even things up on both sides. An enzyme called ATP synthase uses this spurt of energy to put a phosphate group onto ADP. In summary, protons continually get pumped across the membrane, flow backward, release some energy, and thus keep your cells supplied with the ATP you need.

Oxygen

To live, we know we must inhale oxygen continually and consistently. An adult human consumes about a half-gallon of oxygen every minute. You can think of oxygen as the most essential of all nutrients.

Oxygen dissolves poorly in fluids such as blood. This makes things difficult for your circulatory system, which must transport oxygen to all tissues. The body keeps oxygen in the blood during circulation by using respiratory pigments. These pigments are proteins that help oxygen dissolve in blood. With the respiratory pigments, a quart of blood carries about 6.7 fluid ounces of blood. Without these pigments, the blood would dissolve only about 0.15 fluid ounce. The main respiratory pigment of humans is hemoglobin.

Oxygen is part of every compound in the 20 combined steps that make up glycolysis and the Krebs cycle. In the ETC, oxygen plays the role of terminal electron acceptor. This means that the oxygen molecule soaks up electrons at the very last step in the ETC. Other organisms on Earth can use molecules other than oxygen as their terminal electron acceptor. For example, some bacteria use metals. Many yeasts use organic compounds to accept electrons and then form new organic compounds, such as alcohol. But no organism produces as much energy as efficiently as aerobic respiring organisms that depend on oxygen for life.

Respiration's ATP Output

The goal of all energy-generating pathways is to make ATP. With this source of readily usable energy, the body can power its maintenance tasks, repair injuries, digest more food, and move and think. Without energy, it all stops.

Let's sum up the ATP output from each part of your energy-generating system:

➤ Glycolysis produces two ATPs from each glucose.

➤ The Krebs cycle produces two ATPs from each revolution starting with acetyl CoA.

➤ The ETC produces 34 ATPs from each set of electrons released by glycolysis and the Krebs cycle.

➤ The cell must spend about two ATPs to transport the end products of glycolysis from the cytoplasm to the inside of the mitochondrion.

The total ATP production by respiration is about 32 ATPs. Some inefficiencies exist with all this ATP generation, so this number is probably more theoretical than actual. Nevertheless, you will soon see that respiration is still far more efficient for producing energy than other types of metabolism.

CHAPTER 9

Fermentation

In This Chapter

➤ The basics of fermentation and the main types

➤ How microbes get energy without using oxygen

➤ Types of fermentation important to you

In this chapter you will see why fermentation is an important type of metabolism in biology (for reasons other than making wine!) Fermentation is fascinating because it gives energy to organisms that couldn't care less whether oxygen is available in the atmosphere. These organisms are called anaerobes.

The word "aerobic" refers to any situation in which oxygen is present. The word "anaerobic" means "without oxygen." The organisms that use fermentation live in places where oxygen has been depleted from the surroundings. Look for anaerobic habitats in nature in the sediments and mud of swamps, lagoons, and still ponds. Oxygen-depleted places also occur in small spaces between particles of soil as well as the inside of the digestive tract of animals and insects.

Your intestines is one of nature's anaerobic habitats and is home to many hundreds of species of anaerobic bacteria and protozoa.

Anaerobic organisms that ferment organic compounds are all microbes: archaea, bacteria, protozoa, and yeasts. Because they are microscopic, these microbes can carve out an existence in some places where you would think oxygen deters their growth. In addition to living in mud and in the gut, anaerobes grow on your skin and in your mouth. They also live in water distribution pipes, in food, and can cause serious infections in wounds. They survive in these places by finding tiny oxygenless pockets called *microenvironments*.

The species that manage to survive without oxygen serve a variety of useful purposes on Earth. This chapter shows how anaerobic microbes that use fermentation decompose our wastes, help digest our food, and make useful products.

Getting Energy without Using Oxygen

Humans and other respiring species oxidize their food with oxygen. To oxidize a molecule means to put oxygen onto it in a chemical reaction. In this step, respiring organisms take the chemical energy out of the food and convert it to another form of energy: work energy. The work energy runs all the systems that keep you alive. Without oxygen, respiring organisms would disappear from the planet.

For the chemists among you, oxygen plays its vital role in respiration by acting as a terminal electron acceptor. Terminal electron acceptors work at the final step of the electron transport chain. In fermentation, energy generation happens in a similar manner; electrons released in the chemical breakdown of food must be disposed of so the reactions can continue. Fermentation, however, uses organic compounds as the electron acceptors rather than oxygen.

In summary, the two major non-photosynthetic energy processes in nature are:

➤   Respiration—Uses oxygen as the terminal electron acceptor and is present in aerobic microbes and all higher animals.

➤   Fermentation—Uses organic compounds as the terminal electron acceptors and is present in anaerobic prokaryotes, protozoa, and yeasts.

Fermentation gives an advantage to anaerobic microbes when no oxygen is available. These species thrive where respiring organisms cannot. A fermentative organism can survive by using foods that other organisms cannot use for energy. For example, fermentative species digest the plant fiber cellulose, which animals cannot digest. Some fermentative species also use alcohol, lactic acid, and other organic acids for energy. Despite what some people might tell you, getting by on alcohol alone to meet your energy needs does not work for humans!

Fermentation starts with one organic compound, such as a sugar, and ends with a different organic compound:

Substrate A + ADP → Product B + ATP (Energy)

Fermentation begins exactly as respiration begins, with glycolysis. But after producing two pyruvates in glycolysis, the metabolisms of aerobic organisms and anaerobic organisms diverge.

But fermentation has a drawback. It produces less usable energy in the form of ATP than does respiration. Respiration yields almost 20 times more ATP than fermentation.

Biology Tidbits

Glycolysis and Evolution

Glycolysis plays a central role in both respiration and fermentation. Ancient prokaryotes depended on glycolysis as an energy-generating pathway during a time when the Earth's atmosphere contained almost no oxygen. The steps of glycolysis do not require oxygen, so prokaryotes could generate the energy they needed to live without concerns about oxygen levels in the air. The Earth's atmosphere began accumulating oxygen about 2.7 billion years ago, but the earliest prokaryotic fossils date to 3.5 billion years ago. Thus, glycolysis is likely a remnant of the earliest organisms to develop on this planet. Glycolysis remains a central metabolic pathway today in almost every living thing. For this reason, glycolysis has been called biology's "universal pathway."

Most fermentations also produce carbon dioxide (CO_2) as an end product. (Respiration also produces this gas.) To prevent this gas from building up and shutting down further fermentation, many anaerobes depend on a specialized group of microbes, called *methanogens*, to remove the excess carbon dioxide. They do this by converting the carbon dioxide to the gas methane (CH_4).

Biology Tidbits

Methane

The bubbles coming from the depths of swamps and dark lagoons carry gases produced by fermentation, mainly carbon dioxide. Often, a diverse group of archaea and bacteria live in these forbidding habitats and gobble up carbon dioxide for energy and carbon. These microbes called methanogens use carbon dioxide to accept excess electrons as respiring organisms use oxygen. The methanogens produce methane as an end product of their metabolism.

Methane rises from places such as sediments, muds, and landfills, and it is also released continually by *ruminant* animals as a normal product of fiber digestion. Methane enters the atmosphere where, like carbon dioxide, it helps trap some of the Sun's heat and thus warm the planet. This warming is critical for life on Earth. We now know that too much of these gases causes an increase in atmospheric temperatures known as global warming.

Types of Fermentation

About 15 different types of fermentation have been studied by microbiologsts. Many other types of fermentation may exist in nature that have not yet been discovered. Some of the fermentations that are important in nature or for commercial reasons are:

➤ Acetone-butanol Fermentation—Used for making the industrial solvents acetone and butanol by fermenting glucose by *Clostridium* bacteria.

➤ Alcohol Fermentation—Used for making ethanol from sugars in winemaking and other spirits, mainly by the yeast *Saccharomyces* and *Zymomonas* bacteria. This fermentation is also the reason for food spoilage, such as that of fruit juices and ciders.

➤ Butanediol Fermentation—Not commercially valuable, but the ability to carry out this type of fermentation helps microbiologists identify certain microbial species in the environment, especially fecal microbes that have contaminated food and water. The products from glucose are ethanol, 2,3-butanediol, and formic acid.

➤ Butyric Acid Fermentation—Important in the digestion of dietary fiber, this fermentation is performed by bacteria of the digestive tract.

➤ Dihydroxyacetone Fermentation—Used for making the suntan product from glycerol using the bacterium *Gluconobacter*.

➤ Homoacetate Fermentation—Used for making acetic acid from various sugars by a variety of bacteria.

➤ Lactic Acid Fermentation—Used for making lactic acid from various sugars by various bacteria, including *Lactobacillus* and *Streptococcus*. This is also an important pathway in the spoilage of many foods.

➤ Malolactic Fermentation—Important in the preparation of foods such as pickles and soy sauce, *Lactobacillus* bacteria convert malic acid to lactic acid.

➤ Mixed Acid Fermentation—Carried out by various bacteria of the digestive tract, this pathway converts sugars to a variety of acids and alcohols.

➤ Propionic Acid Fermentation—An important component of fiber digestion in the gut of cattle and sheep, various bacteria convert glucose and lactic acid to the organic acids propionate and acetate.

➤ Sorbose Fermentation—Used by manufacturers of ascorbic acid (vitamin C) by allowing the bacteria *Acetobacter* or *Gluconobacter* to convert sorbitol to a vitamin precursor.

Although discussions of fermentation could fill a book and make your head spin, let's look at only a few types that are most important to you.

Alcohol Fermentation

Alcohol fermentation is a two-step process by which pyruvate from glycolysis is turned into ethanol. People have known about this fermentation since the earliest civilizations. It is used in winemaking, brewing, baking, and many other food preparations.

Various bacteria carry out alcohol fermentations, but the best-known organism harnessed by the food and alcohol industries is the yeast *Saccharomyces*.

Biology Tidbits

Saccharomyces

Saccharomyces is nicknamed baker's yeast because of its necessary role in the fermentations used in breadmaking. The bubbles in bread result from the carbon dioxide emitted by this microbe when it ferments sugars. In the making of fermented beverages, ethanol is the main desired end product rather than carbon dioxide.

Lactic Acid Fermentation

Lactic acid fermentation is so-named because certain bacteria and fungi that use this pathway produce mainly lactic acid from the pyruvate from glycolysis. ("Lactate" is another word for lactic acid.) Some microbes produce nothing but lactic acid. This type of fermentation is called homolactic fermentation. Other microbes produce one or two compounds in addition to lactic acid. This is called heterolactic fermentation.

Biology Vocabulary

Lactate

When the body's muscles must work at a rate faster than oxygen intake can keep up, such as strenuous exercise, the muscles switch over to lactic acid fermentation. Lactate begins to accumulate in the muscles. You know this as muscle fatigue. If enough lactate accumulates, the lactate causes temporary pain in the muscles. The blood gradually carries away the excess lactate and delivers it to the liver, which converts it back to pyruvate.

The terms "aerobic exercise" and "anaerobic exercise" relate directly to lactate metabolism in your body. Aerobic exercise means that your body's oxygen uptake and respiration are keeping pace with your muscles' work. Anaerobic exercise is usually associated with a sudden burst of intense activity in which the muscles call upon fermentation to keep pace with your activity.

Mixed Acid Fermentation

The tender steak, juicy hamburger, or cold glass of milk you might be envisioning right now are products of mixed acid fermentations. These fermentations take place in the large stomach of ruminant animals and are performed by hundreds of different species of bacteria. A variety of organic acids result from these fermentations plus ethanol, carbon dioxide, and hydrogen.

To keep the end products from building up, the rumen also contains bacteria that absorb specific end products of mixed acid fermentation and use them for their own energy needs. But some of the acids escape the rumen and go to the animal's intestines, where they are absorbed into the bloodstream. The ruminant then uses the organic acids in its energy metabolism, for building new muscle, and in producing milk.

The specific acids absorbed by ruminants give milk their distinctive flavors. For this reason, cow's milk tastes different than goat's milk. Even the milk from certain cow breeds differ in flavor and color, in part because of the organic acids produced by bacteria in the animal's rumen.

Biology Tidbits

Digestion in Ruminant Animals

Fermentation of dietary fiber in the stomach of cattle, sheep, and goats dominates all other types of microbial metabolism in this organ. The largest section of a ruminant's stomach, called the rumen, has in fact been likened to a fermentation vat and contains billions of fiber-digesting bacteria and protozoa.

Few animals digest fiber as efficiently as ruminants (cattle, sheep, goats, deer, elk, elephants, and giraffes). Horses and rabbits are also fairly good at fiber digestion by using an enlarged digestive organ called the cecum. Fiber-eating insects such as termites and cockroaches also break down woody fibers because their digestive tracts contain many of the same bacteria as found in the rumen.

Microbial enzymes in these species break down plant fiber to glucose. But unlike humans that absorb the glucose directly and use it for energy, ruminants ferment the glucose to organic compounds called volatile fatty acids (VFAs). Once absorbed into the bloodstream, these VFAs play an important role in the ruminant animal's energy metabolism. The VFA acetic acid enters the Krebs cycle. A second VFA, propionic acid, is converted in the liver to glucose. The animal then burns this glucose via glycolysis.

Digestion in meat-producing ruminants is important to us because it is a major way of converting the energy in plants that are poorly digested by us into a more digestible food: meat.

CHAPTER 10

Photosynthesis

In This Chapter

➤ The two stages of photosynthesis and their purpose

➤ The Earth's photosynthetic organisms and their features

➤ The way photosynthetic pigments work

➤ How carbon dioxide and light is absorbed and oxygen is made

In this chapter we will explore photosynthesis. By this natural process, certain organisms turn solar energy into chemical energy that can be used by humans and other animals.

Photosynthesis is the third major type of energy-generating metabolism on Earth along with respiration and fermentation. The respiring and fermenting organisms could not survive, however, without the organisms that perform photosynthesis.

Photosynthesis provides three life-giving services to Earth. First, as mentioned, it transforms the Sun's energy into a form we can use. Second, it replenishes the atmosphere with oxygen. Finally, photosynthesis removes carbon dioxide from the atmosphere so that it doesn't build up to lethal levels. Animal life simply cannot survive without the plants and microbes that perform photosynthesis. No wonder it has been called the biochemical process that feeds Earth.

Organisms that can perform photosynthesis contain organelles called chloroplasts. The chloroplasts act as the site where the cell puts carbon dioxide molecules together to make glucose. This glucose then provides all of the organism's energy needs plus the carbon for building all other constituents.

Plants use about 50 percent of the glucose from photosynthesis for respiration inside their mitochondria. The remaining 50 percent goes to making new plant fiber, starch, proteins, and other components of the plant body. Look at photosynthesis this way: though we cannot live without plant life, photosynthetic plants can do fine without animal life.

Photosynthetic Organisms

Photosynthetic organisms survive without needing organic compounds as food because they make their own glucose from carbon dioxide. These organisms are called autotrophs. By contrast, all animal life belongs to a second group based on nutrition. We are heterotrophs. Heterotrophs require organic compounds to live.

The photosynthetic autotrophs range from microscopic bacteria to huge redwoods. The Earth's autotrophs are:

> ➤ Green plants and trees.

> ➤ Algae.

> ➤ Some protists, such as *Euglena*.

> ➤ Cyanobacteria.

> ➤ Purple sulfur- and nonsulfur-using bacteria.

> ➤ Green sulfur- and nonsulfur-using bacteria.

Biology Vocabulary

Heterotrophs and Autotrophs

The Earth's living things can be divided into two groups based on their nutrition. Autotrophs can feed themselves by absorbing energy from the Sun and carbon from the atmosphere. Heterotrophs must depend on taking in organic compounds for their energy and carbon. Because autotrophs supply energy to all other life on Earth by first capturing the Sun's energy, ecologists call these organisms "producers." By contrast, heterotrophs are called "consumers" because they subsist by consuming other biological things.

Two types of autotrophs exist. Organisms that use sunlight as their energy source are called photoautotrophs. Specialized prokaryotes that get energy from reactions involving inorganic chemicals, such as metals, are called lithoautotrophs.

Although plants, trees, algae, and protists have vast diversity in size and structure, they all have in common the presence of chloroplasts in their cells. Prokaryotic cyanobacteria and the other photosynthetic bacteria, by contrast, lack this organelle.

Algae and cyanobacteria make up a group of organisms known as phytoplankton. These organisms play a monumental role in Earth ecology. They possibly put more oxygen into the air and absorb more carbon dioxide than all terrestrial plants and forests.

Biology Vocabulary

Phytoplankton

Plankton is a general name for all the tiny to microscopic organisms that live in surface waters and serve as food for other organisms. Plankton can be divided into plant-like organisms (phytoplankton) and animal-like organisms (zooplankton). Phytoplankton includes mainly algae, cyanobacteria, and tiny aquatic plants. Zooplankton contains nonphotosynthetic organisms such as protozoa and invertebrates. All plankton serves as food for other living things.

Phytoplankton occupies the upper layers of freshwaters and marine waters exposed to sunlight. In most of these places, the mass of total phytoplankton can be tens to hundreds times greater than the mass of zooplankton and other life forms. Phytoplankton supports all the other life by being a *primary producer*. Primary producers are the organisms on Earth that first capture the Sun's energy and turn it into a form usable by all other life.

Cyanobacteria

This group of bacteria deserves special mention because of its importance in photosynthesis today as well as in cyanobacteria's role in the evolution of respiring organisms.

Modern plant life began evolving from photosynthetic algae about two billion years ago. Before then, the atmosphere already contained some oxygen. This oxygen had been supplied by some of Earth's most ancient microbes, the cyanobacteria. Cyanobacteria may have developed as early as 3.5 billion years ago. By living on atmospheric carbon dioxide and developing the unique ability to absorb energy from sunlight, they held a special place among the earliest prokaryotes. Rather than being killed by the toxic effects of oxygen, cyanobacteria tolerated the gradual buildup of oxygen in the atmosphere. The cyanobacteria were, after all, solely responsible for this buildup. Billions of years would pass before the oxygen levels reached that of today's atmosphere, about 21 percent. As photosynthetic eukaryotes evolved, the oxygen rise accelerated, but we owe our current life-giving conditions on Earth to the first oxygen-producing prokaryotes.

Biology Vocabulary

Purple and Green Bacteria

Purple bacteria and green bacteria are photosynthetic microbes that are so-called because of their pigmentation. Purple bacteria actually range in color in their water habitats from pink to reddish purple to brown. Green bacteria range from green to yellowish brown.

All purple and green bacteria species can be further divided into groups that use sulfur as part of their energy metabolism and those that do not need sulfur. The sulfur-using species can sometimes be found in nature by the yellow sulfur granules they excrete, which build up until you can see them without needing a microscopic.

Purple and green bacteria differ from other photosynthetic organisms by an important factor: they do not produce oxygen. This type of photosynthesis is called anoxygenic photosynthesis. It may be an ancient form of metabolism that predates even the oxygen-producing cyanobacteria.

Cyanobacteria live in both water habitats and on land. Their only need is to be near the surface of water, soil, rocks, or other surfaces so that they stay exposed to sunlight. Cyanobacteria often form thick mats in freshwaters, in hot springs, and in some marine waters. They also cover boulders, tree trunks, and even house roofs—any light-exposed surface that also offers some moisture can usually support cyanobacteria.

These microbes come in diverse forms from single cells to long strands called filaments. When billions of filamentous cyanobacteria congregate on the water's surface, we sometimes give them the uncomplimentary name "pond scum." Many cyanobacteria contain pigments that impart a bluish green color. When they were first studied in microbiology, scientists incorrectly classified them as algae because of their coloring and called them "blue-green algae."

Chloroplasts

The chloroplast is Photosynthesis Central for eukaryotic cells. Two membranes enclose this elegant structure. A narrow space separates the outer membrane from the inner membrane. The inner membrane holds a dense fluid called stroma, which is filled with interconnected sacs called thylakoids. The thylakoids occur in stacks like poker chips, but if we were to see a cross section of a thylakoid, we would see that each disc-shaped sac connects to the one above it in the stack and the one below. Biologists call these stacks or columns grana.

The pigment chlorophyll fills the thylakoid sacs in photosynthetic organisms that produce oxygen. Thus, it is inside the thylakoids where photosynthesis takes place. Photosynthetic pigments play the essential role of absorbing light to begin the process.

Light and Photosynthetic Pigments

The pigments of photosynthetic organisms serve as the material that absorbs light. Physicists refer to light as electromagnetic energy or electromagnetic radiation. This form of energy is composed of waves, and all light can be defined by its wavelength.

If you were to picture a wave moving across a small pond, you could spot the uppermost part of each wave, called the apex, and the bottom of each wave, called the trough. The distance from apex to apex (or trough to trough) is one wavelength. The entire range of all light's wavelengths makes up the electromagnetic spectrum. Most living organisms are most interested in the part of this spectrum from a wavelength of 380 nanometers (nm) to 750 nm. (A nm is one-billionth of a meter or about 0.00000004 inch.) This part of the radiation spectrum is called visible light.

Pigments are materials that produce color. They do this by absorbing light in a particular narrow range of the spectrum. Many pigments exist in nature and scientists can differentiate most pigments from each other by using an instrument that shows the wavelength a pigment absorbs. This instrument is a spectrophotometer. For example, various chlorophylls absorb only light in wavelengths that produce violet-blue light (around 450 nm) and also red light (around 650 nm.)

Different photosynthetic organisms use different pigments for absorbing light. The main natural pigments involved in photosynthesis are:

➤ Chlorophyll *a*—Absorbs violet-blue and red light and occurs in most photosynthetic eukaryotes; probably the main photosynthetic pigment on Earth. Because chlorophyll *a* absorbs these wavelengths, it appears to us as bluish green.

➤ Chlorophyll *b*—Absorbs blue light and red light at wavelengths slightly longer than chlorophyll *a*'s range. Appears to us as yellow-green.

➤ Carotenoids—Various pigments that absorb violet and blue-green light. They play a lesser role in photosynthesis but may protect the chlorophylls from damage by absorbing excess light. Their appearance ranges from yellows to oranges.

> ➤ Bacteriochlorophylls *a* and *b*—Used by the purple and green bacteria, these pigments absorb long wavelength light called infrared. The bacteriochlorophylls enable the bacteria to live in deeper water than the organisms that must live at the water's surface to absorb direct light.

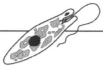

Inside a Cell

Chlorophyll

The pigment chlorophyll is almost synonymous with photosynthesis, and rightly so. The green plants, algae, and cyanobacteria that dominate the Earth's oxygen production all use this pigment to capture light for photosynthesis. You know chlorophyll as the pigment that gives plants and algae their green color.

When sunlight hits a chlorophyll molecule, the light energy causes chlorophyll's electrons to move faster than normal. Chemists call this the molecule's "excited state." In a split second, excited electrons change the energy level in one chlorophyll molecule to the next in a miniature chain reaction. A specialized region of the chlorophyll molecule captures the energy given off by excited electrons.

After chlorophyll captures light energy, it uses it to make ATP from ADP. The ATP in turn powers a metabolic route that builds sugar out of carbon dioxide units. When chlorophyll is studied in a test tube, the light it absorbs is disconnected from normal photosynthesis. In this situation, the pigment fluoresces when exposed to light.

Photosynthesis in Bacteria

Bacteria lack organelles so therefore they lack chloroplasts. Photosynthetic bacteria instead have regions where their plasma membrane folds repeatedly upon itself. As a result, the membrane forms stacks of these folds that resemble the thylakoid inside chloroplasts. Named "thylakoid membranes," these regions work the same way as true thylakoids of chloroplasts.

Photosynthesis' Two Stages

Photosynthesis can be a complicated process to understand. It's made even more complex by various alternate routes used by many cells for shuttling excited electrons from pigment to pigment. A major part of the system behaves similarly to chemiosmosis you learned about in the previous chapter for making ATP. I will present photosynthesis here at its most basic by describing its two stages. Biologists have (thankfully) given these stages easy names: the "light reactions" and the "dark reactions."

The light reactions of photosynthesis are involved in capturing light and transforming it to a usable form of energy. The dark reactions of photosynthesis concern themselves with using that energy to make glucose.

Light Reactions

The chemical reactions of photosynthesis could make chemiosmosis appear almost elementary. Let's consider only the main events of the light reactions, all of which occur in the thylakoid's membrane:

> ➤ Light strikes a pigment, causing the pigment's electrons to change to an excited state.

> ➤ A pigment-protein complex captures the energy from the excited electrons and absorbs a water molecule.

> ➤ The water splits to produce hydrogen used for the energy-generating steps and releases oxygen.

> ➤ The pigment cytochrome transfers the electrons to a second photosystem while pumping protons across the thylakoid membrane.

> ➤ The resulting proton gradient drives the formation of ATP.

> ➤ The electrons shuttle to the dark reactions where they are used in glucose synthesis.

Biology Tidbits

Fluorescence

Many pigments have the ability to fluoresce, meaning they glow for a period of time after being exposed to light. When a pigment absorbs light, some of the pigment's electrons change from their normal ground state to an excited state. If the pigment is not part of a photosynthetic system (for instance, a chlorophyll solution in a test tube), the electrons rapidly return to their ground state. When this happens, the electrons give off a bit of energy as light plus a little heat. We observe this light emission as an afterglow that lingers for a second or two.

Dark Reactions

As you might have guessed, the dark reactions of photosynthesis take place even if no light is present. This stage of photosynthesis uses the energy from the light reactions to make new sugar molecules, which then fuel every other reaction in the cell. (The dark reactions contribute the "-synthesis" part of photosynthesis.)

The dark reactions represent a kind of glycolysis in reverse. In this process, the cell consumes carbon dioxide and builds these two-carbon units into bigger molecules until it makes a six-carbon sugar (glucose). This metabolic pathway is called the Calvin cycle. The seven steps of the cycle were first described in the 1950s by the American chemist Melvin Calvin with help from biologists Andrew Benson and James Bassham.

Every time a carbon dioxide enters the Calvin cycle, this step is called carbon fixation. In bacteria, carbon fixation occurs when the molecule simply diffuses across the cell membrane. Plants use a more sophisticated process for carbon fixation. The underside of plant leaves contains hundreds of tiny pores called stomata. The stomata open and close to take in carbon dioxide from the air and release oxygen. In this way, the stomata keep the dark reactions of photosynthesis supplied with carbon and allow the plant's end product, oxygen, to escape.

PART THREE

Basics of Genetics

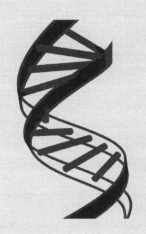

CHAPTER 11

Deoxyribonucleic Acid (DNA) and Chromosomes

In This Chapter

➤ A description of your cells' DNA and its replication

➤ The basics of genes and chromosomes

➤ How chromosomes divide and determine a newborn's traits

In this chapter you will move on to processes involved in making new cells just like the original cell. A large part of the metabolism you learned about in preceding chapters relates to the body's need to grow, reproduce, repair injury, and replace worn-out components. Good nutrient supply and metabolism running at its best help make these things happen. How does your body's cells know what to do? How do they know when to turn on certain parts of metabolism and shut down other parts for awhile? How do they know when to make new cells?

All of the information passed down from your ancestors and available to your offspring and future generations resides in your DNA. Let us begin exploring the molecule that makes you who you are.

Your Cells' DNA

All of the information critical for running your body resides in the famous material known as DNA. DNA is a molecule called a nucleic acid. The full name deoxyribonucleic acid tells a little about its composition. It is a nucleic acid containing ribose ("-ribo-") that lacks an oxygen atom ("deoxy-") that is normally part of this sugar. As you learned in Chapter 2, a nucleic acid contains a long string of sugars plus nitrogen-containing bases plus phosphate groups.

In DNA, the deoxyriboses link to each other to make two long strands. Phosphate groups stick out from each sugar. The nitrogen base attaches to the sugar at the opposite side from where phosphate attaches. The two strands of DNA connect to each other by chemical bonds between the protruding base from one strand and the protruding base from the second strand. When the two strands so connect, they form a double strand resembling a ladder. Because this double strand tends to twist a bit into a helical shape, DNA's form is often called a double helix.

Biology Tidbits

Watson and Crick and the Double Helix

When DNA's double helix structure was discovered in the 1950s, it became known as the Watson-Crick model. American James Watson and British Frances Crick, two fairly unknown scientists at the time, unraveled the three-dimensional chemical structure of DNA. Though Watson and Crick entered history as the discoverers of DNA's structure, they could not have reached this milestone without help from physicist Maurice Wilkins and his colleague Rosalind Franklin. Wilkins and Franklin used a technology called X-ray crystallography to make black and white images of the DNA molecule. These images led Watson and Crick to surmise that the substance had a helical shape.

Watson and Crick began building models of helical strands at their desks. By pairing and re-pairing the bases connecting both DNA strands, they devised a structure that fit together and would match the results from Wilkins and Franklin's lab. The specific base-pairing also turned out to be essential in the genetic code held within DNA.

DNA seems to be among the most mysterious substances in nature because of the vast information it holds. Yet you may be surprised to know that any high school biology student can get DNA out of cells in a few easy steps. The molecule can be coaxed from a salt solution by gently swirling a glass rod in the liquid. A sticky blob of colorless DNA clings to the rod like a genetic lollipop.

Biologists' fascination with DNA resides in the base pairs that connect DNA's strands. These pairs produce what biologists call "complementary strands" of DNA.

Complementary Strands

DNA the world over contains only four different bases. These bases are adenine, thymine, guanine, and cytosine. Scientists abbreviate them to A, T, G, and C, respectively.

Adenine always pairs with thymine, and only thymine. Guanine similarly pairs only with cytosine. The rungs of DNA's ladder therefore consist of thousands of A-T, T-A, G-C, and C-G pairings. This is why the two strands are said to be complementary to each other. By knowing the sequence of bases on one strand, you automatically know the corresponding sequence on the other strand.

Take any single strand and decipher the base sequence. A scientist might, for instance, find a section of a strand that looks like this: A A T T A G C T A T T T G C G C G A A A. Can you fill in the base sequence for the complementary strand? Try it. (The answer is: T T A A C G A T A A A C G C G T T T.)

The Genetic Alphabet

The four letters corresponding to the four bases of DNA make up biology's genetic alphabet. How can only four letters carry all the information that makes you who you are, distinguish you from your siblings, and make you a human rather than a slug? Biology does this by using the bases in groups of three, called triplets or *codons.*

The different possible combinations of letters in a codon greatly expands the language spoken by DNA. Each different arrangement of three consecutive bases gives 64 possible "words." These words make up the *genetic code.* That is, they carry all the information pertaining to who you are, how you evolved, and who your offspring will be.

What Is a Gene?

DNA contains genes, which you can think of as sentences of the genetic code formed by the words or codons. Genes can be made of several codons to several dozen codons. They occupy specific sites along the DNA molecule and must always contain the bases in a set order. Even one incorrect base in the sequence can change drastically the information carried by a gene.

Biology Vocabulary

Genotype and Phenotype

Every living thing has a genotype and a phenotype. Your genotype is invisible to you and to others. It resides in your genetic makeup or, specifically, your exact genetic code. By contrast, your phenotype represents the physiology and physical traits that are determined by your genotype. Hair color, eye color, ability to metabolize drugs, and how fast you run are but a few examples of phenotype.

In the next few chapters, you will see how genes run all of the body's operations. Your exact gene makeup represents your genotype. As you will also soon see, the body uses the information contained in genes to affect physiology and outward appearance. These measurable characteristics that result from the information carried in genes are called your phenotype.

Surely a caring, feeling human holds more genes than any other living thing, right? A human has almost 25,000 genes. But size, intelligence, and other clues to a being's complexity seem to relate little to its number of genes. See the variety of gene numbers (approximate) in some of Earth's inhabitants:

> ➤ The water flea *Daphnia pulex* contains 31,000 genes, highest now known among all animal life.

> ➤ *Escherichia coli* (*E. coli*) bacteria contain 3,200.

> ➤ The roundworm *Caenorhabditis elegans* contains 19,000.

> ➤ The house mouse *Mus musculus* contains 25,000.

> ➤ The rice variety *Oryza sativa* contains 51,000 genes, among the highest known for all life.

> ➤ The protozoan *Trichomonas vaginalis*, which causes sexually transmitted infection, contains 60,000, among the highest known for all life.

It takes more genes to run a mouse than to run a human! Obviously, number of genes does not alone determine an organism's complexity. Complexity in an organism comes from how genes are regulated, that is, turned on and off and the manner by which they coordinate with other genes.

Biology Tidbits

Transposons

Bacteria contain sequences in their DNA that can lift off and move to another location in the DNA molecule. These sequences are called transposons. Transposons caused considerable debate among geneticists when they were discovered by biochemist Barbara McClintock in the 1940s. Most scientists doubted a functioning part of DNA could move around without killing the cell. Transposons are now known to help bacteria adapt to new environments, but this concept had been so hard for scientists to accept, McClintock would wait forty years before being recognized for her discovery. She was awarded the Nobel Prize in Physiology in 1983.

How DNA Replicates

When cells replicate to make two new cells, they must also replicate the DNA exactly so that the mistake-free genetic code goes to the new offspring. This is true for prokaryotic cells that simply split in two as well as eukaryotic cells that use a more complex type of reproduction.

DNA replication must be correct every time a cell divides. Mistakes in the replication steps can lead to genetic disorders, disease, and even death in some organisms.

Replication of the DNA molecule starts at certain specific sites called origins of replication. A single DNA molecule can hold several of these origins so that, rather than go from one end to the opposite end, DNA starts to replicate at several different spots simultaneously. At these sites, the two strands come apart to form a bubble-like bulge in the DNA. An enzyme jumps into the bubble and begins working on the unattached sections.

The enzyme (called DNA polymerase) begins attaching complementary bases to each base on each unattached DNA strand. As you already learned, A always connects with T and G always binds with C. As the bubble moves along the molecule, DNA polymerase continues putting bases onto each strand. From one double helix molecule grows two new double helices.

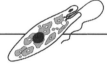

Inside a Cell

DNA Polymerase

The enzyme DNA polymerase controls the replication of DNA in all living things. Although I use the term to suggest a single enzyme is at work, the term "DNA polymerase" refers to a group of enzymes that coordinate their activities. For example, in bacteria such as *E. coli*, two DNA polymerases perform the replication of DNA. In many eukaryotes, more than ten different DNA polymerases do the job. Regardless of the number of enzymes used, they work the same. They start at an origin of replication, proceed along a detached section of the DNA ladder, and produce two new copies of the original DNA.

When one strand of DNA detaches from its complementary strand, each single strand represents the blueprint for the organism's genetic code. Biologists call this strand the *template*. DNA polymerase reads the bases in the template and selects the correct corresponding base to attach onto it. When the two strands separate, they are different molecules. Two features of DNA replication ensure that the final product will be an exact replica of the original double strand:

➤ Correct and complementary base parings.

➤ Antiparallel orientation of the two strands, meaning that DNA polymerase reads one strand from east to west and the other strand from west to east.

A small handful of other enzymes help out during DNA replication. These enzymes do a kind of "clean up" similar to brushing away loose pieces of thread or fabric when you've stitched up a tear in your jeans. Important enzymes in DNA replication are:

➤ Helicase—Unwinds the double helix prior to replication.

➤ Topoisomerase—Corrects overwinding during replication.

➤ Primase—Starts the replication process by building a primer that is complementary to a short segment of one strand.

➤ DNA Ligase—Joins any fragments made during the replication.

Each of these enzymes have more intricate responsibilities than I'll discuss here, but the basic principles of DNA replication remain the same whether we talk about fruit flies or zebras.

Chromosomes

The chromosome resides inside the nucleus of eukaryotic cells. Thus, the nuclei of your cells contain all your genes.

In eukaryotic cells, a chromosome is a long threadlike piece of DNA plus certain proteins that protect and support the DNA. Many species distribute their entire DNA content among several chromosomes. (Prokaryotes do not use many proteins to support their DNA molecule; in prokaryotes, DNA and chromosome are interchangeable terms.) As the list below shows, different species vary by their number of chromosomes:

➤ Human, 46 chromosomes.

➤ Dog, 78.

➤ Crayfish, 200.

➤ Yeast, 32.

➤ Ant, 2.

➤ Horsetail plant, 216.

Most of a human's DNA is packaged into 23 pairs of chromosomes in the nucleus. All cells except red blood cells carry chromosomes. One set of the 23 pairs came from your father and the other set came from your mother. The only exception to this pattern exists in reproductive cells. Sperm cells and egg cells contain only one set of 23 chromosomes. When the two combine during reproduction, they each contribute their half-set to produce a full set of chromosomes.

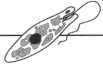

Inside a Cell

Chromosome Number

The 23 pairs of chromosomes in humans add up to 46 chromosomes. This is called the diploid number of chromosomes. Put another way, the normal number of chromosomes for any species is its diploid number. When a youngster grows or when you repair an injury, the body makes new cells. All the cells of the body replicate the chromosomes so that the new cells will also receive a diploid number of chromosomes.

The only exception to this process happens in the reproductive organs. Females and males make new reproductive cells that carry only half of the diploid set of chromosomes. Biologists call this a haploid number. When female and male reproductive cells, each carrying a haploid number of chromosomes, combine during *fertilization*, a new cell is made. This new cell again holds the full diploid set of chromosomes.

Genes are located on specific sites on the chromosomes. Many of these genes control metabolic functions in your body of which you remain blissfully unaware every day of your life. Some genes, however, control the characteristics that define you as an individual. These characteristics are called *traits*. Reproductive cells play a central role in determining the traits you inherit from your parents.

How Chromosomes Control the Traits You Inherit

Both chromosomes and genes occur in pairs in the nonreproductive cells that make up your body. Biologists call each paired set of chromosomes *homologous chromosomes*. The corresponding paired genes at the same site on two different homologous chromosomes are called *alleles*.

Everyone becomes a unique individual due to subtle differences between two alleles. For example, a set of alleles on homologous chromosomes may determine eye color. But you have blue eyes and your brother has brown eyes. How could that happen if you have the same parents and should be getting the same traits from both at time of fertilization? The answer comes from reproductive cells, alleles, and the manner in which chromosomes divide during a process called meiosis.

During cell replication, the chromosomes duplicate to form two identical copies of each chromosome. When cells divide, each new daughter cell—they are never called "sons"!— gets a full set of chromosomes, the diploid set. In reproductive cells, further chromosome division takes place. In these steps, certain things happen that can increase the genetic variations given to a new generation:

➤ Meiosis gives half the homologous chromosomes to one set of reproductive cells (gametes) and the other half to a second set.

➤ During meiosis, some homologous chromosomes exchange short segments with each other. This genetic rearrangement of alleles is called crossing over.

➤ At the end of meiosis, each sperm or egg cell receives one-half of the chromosomes. These cells are haploid cells.

➤ Crossing over leads to sperm and egg cells that carry the same genes but different alleles in almost limitless combinations.

➤ The haploid cells merge their respective half-sets of chromosomes during fertilization.

➤ The specific sperm and egg that merge in fertilization is a random event. This randomness leads to almost limitless variations in the alleles to be inherited from the father and the mother.

Biology Vocabulary

Meiosis

Meiosis is a process that occurs only in the ovaries of females to produce eggs or the testes of males to produce sperm cells. Meiosis' purpose is to reduce the chromosome number from the normal two sets (46 in humans) to one set (23 in humans).

In a resting phase of meiosis, called interphase, the cell duplicates all its chromosomes.

Meiosis then begins. It consists of eight steps comprising two stages. The first stage separates the two sets of homologous chromosomes inside a reproductive cell and cleaves the cell in two. The second stage separates the chromosome sets and then cleaves the cells again. Meiosis results in four haploid daughter cells.

The combination of crossing over during meiosis and the randomness of events in fertilization lead to genetic variation. Although you are more like your siblings than you are like any other person on Earth, quite a bit of variation exists among siblings because of genetic variation.

In humans and other mammals, only one of the four daughter eggs formed in the ovary fully develops. The other three become halted in their development and do not participate in reproduction. By contrast, in the testes all four sperm cells formed in meiosis develop into active spermatozoa (mature sperm cells).

CHAPTER 12

The Cell Cycle

In This Chapter

➤ Cell division steps in prokaryotes and eukaryotes

➤ The phases of mitosis, or how cells divide

➤ The phases of the cell cycle and why cell cycles matter

➤ The relationship between cell cycles and cancer

In this chapter you will learn about how cell cycles work and why they are important to life.

Everything in the biological world has some sort of rhythm. This rhythm connects an organism to the rhythms of the planet: days and nights as well as seasons. This connection in turn helps an organism find food and water and know the best time for making a new generation of offspring.

Although we may think of cell cycles tied exclusively to cell (or organism) reproduction, this chapter shows why there is much more to cell cycles than just reproduction.

Cell Division

In microbes, plants, and animals, cell division gives organisms a way to reproduce. In all of the Earth's matter, this may be the best way also to distinguish living things from nonliving things. Living things can replicate themselves based on a blueprint held in their genetic material.

In biology, cell division can be of two types: asexually based division and sexually based division.

Cell Division in Prokaryotes

Bacteria and archaea that make up prokaryotes and some protozoa divide by an asexual mode of reproduction. This is called binary fission.

In binary fission, one cell splits into two nearly identical daughter cells. That's it! To make another generation, the daughter cells divide the same way. It continues this way *ad infinitum*.

Microbes can divide easily because they have only one chromosome. The task of replicating the single chromosome once before splitting a cell in two demands little work compared with cell division in eukaryotic cells. All a microbe must be careful to do is be sure that after it replicates its DNA, it separates the copies to each side of the cell. When the cell then divides down the middle, each chromosome copy goes to a new daughter cell.

Cell Division in Eukaryotes

Cell division demands more work from eukaryotes than prokaryotes. Eukaryotes often contain multiple sets of chromosomes. In addition, eukaryotic cells contain membrane-enclosed organelles that require their own replications.

Cell division in eukaryotes probably evolved out of the earlier prokaryotic means of replication. Biologists have indeed indentified proteins from prokaryotic division that also participate in eukaryotic division. Though the exact evolution remains a mystery, the process of eukaryotic cell division called mitosis seems to have evolved out of binary fission.

Mitosis

Mitosis is a five-step process by which a eukaryotic cell prepares its chromosomes for cell division, separates them inside the intact cell, and then cleaves a cell in two. The five steps of mitosis result in six distinct phases of a cell's life cycle:

> ➤ Interphase—The period between cell divisions or any activity leading up to division.

> ➤ Prophase—Loosely arranged DNA of the interphase forms tightly coiled chromosomes and the nucleoli disappear. Some cellular infrastructure associated with cell-splitting starts to appear.

> ➤ Prometaphase—The nucleus' membrane disintegrates and chromosomes attach to a system of *microtubules* that extend from one side of the cell to the other.

> ➤ Metaphase—Chromosomes become more organized along the microtubules and two structures called centrosomes form at opposite ends of the cell, and connect to the microtubules.

➤ Anaphase—The chromosome sets break apart and separate along the microtubules. The two sets slide along the microtubules toward the centrosomes at opposite ends of the cell.

➤ Telophase—The cell cleaves in two and forms two daughter cells, each with a single set of chromosomes. The chromosomes immediately become less dense and the DNA less organized. Part of this phase relates to the separation and equal allocation of cytoplasm. This process is called cytokinesis.

Biology Vocabulary

Centrosomes

Centrosomes consist of material always present in eukaryotic cells, but during mitosis they become more organized. The centrosomes organize and direct the microtubules, which give the chromosomes a framework to follow during cell division. At the end of mitosis, the centrosomes disintegrate and their components disappear back into the cell cytoplasm.

In animals, metaphase is the longest phase in mitosis. This phase can last for 20 minutes. The next phase, anaphase, is the shortest, taking only a few minutes to complete. Unless a body is growing or repairing an injury, cells usually remain the interphase.

Biology Vocabulary

Cytokinesis

Cytokinesis is the division of cytoplasm into the two daughter cells formed in mitosis. This is no small chore considering cytoplasm is mainly water. Imagine trying to divide a water-filled balloon into two new balloons. The cell loses little of its contents during cytokinesis because as the two new cells start to separate, the cell quickly builds a furrow down the middle that holds the contents in each half.

In animal cells, thin filaments help pinch the dividing cell in two and also hold in the cell contents. In plant cells, a rigid plate forms down the middle of a dividing cell. This *cell plate* will become part of the new cell walls of the daughter cells.

The Importance of Cell Cycles

As mentioned, cell division is needed for growing, repairing damage, healing injury, and reproduction. In a larger sense, cell cycles also contribute to the rhythms of a multicellular organism throughout its lifetime.

No living thing can sustain a constant, frenzied state of activity. Cell cycles and the body's cycles help individuals and they also help populations of organisms in several ways:

➤ Provides time to rest, digest food, and store nutrients.

➤ Provides time to repair injury or fight infection.

➤ Gives cells an opportunity to sense changes in their surroundings and respond to them.

➤ Helps with population control by slowing the reproductive rate.

➤ Helps species from overrunning their food supply and habitat.

➤ Enables prey populations to synchronize with predator populations, which keeps Earth's systems in balance (see Chapter 32).

You probably notice that rest and replenishment are as important as reproduction for a species' wellbeing. Mitosis, in fact, takes up only a fraction of a cell's total cell cycle. Most of a cell's time is spent in other activities or in rest.

A Typical Cell Cycle

Biologists have identified distinct phases of a total cell cycle:

➤ M Phase—The period when mitosis occurs. This phase takes up no more than 10 percent of an entire cell cycle.

➤ G1 Phase—A long period of rest or interphase.

➤ S Phase—The period of active synthesis of DNA and other compounds needed in energy storage or for repair.

➤ G2 Phase—Another long period of mainly rest with some preparation for mitosis.

The G phases take up more than 50 percent of a cell's time under normal conditions. The S phase uses about a quarter of the cycle.

How Nature Controls Cell Cycles

Because cell cycles should at times run fast and slow down at other times, every cell has a mechanism for controlling its cycles.

Cells control their cycles like they control almost every other important function: by regulating gene activity. Cells in fact contain specific molecules devoted to triggering and stopping steps in a cell cycle. The controls on a cell cycle come from two equally important sources:

➤ Internal controls—Mainly genes than have evolved specifically to help control cycling as well as the sensors of the cell that tell it when to speed up or slow down.

➤ External controls—Factors in the environment that when detected by a cell induce faster cycling or slower cycling.

Every cell cycle also has at least one checkpoint. A checkpoint is a spot that tells the cycle to either go or stop. Checkpoints help put a brake on runaway cell cycles that would lead to an exhaustion of the cell's food supply, water, and other materials. Animal cells have checkpoints that normally stop the cycle. Only when certain factors are present can the stop command be overridden so that the cycle cranks a few more times.

Most cells possess three main checkpoints and several minor checkpoints, depending on species. The three main checkpoints common in almost all cells are in the middle of the G1 phase, the end of the G2 phase, and midway in the M phase. If a cell does not receive a go-ahead signal from its internal or external controls, then it will not divide or even prepare for cell division. This idling state is sometimes called the G0 phase.

Cancer: When the Cell Cycle Goes Out of Control

A body's cells must reproduce when appropriate and stop reproducing when appropriate. The metabolism that speeds up or slows down in coordination with this cycling must also be controlled. A body in which the cells reproduce and consume nutrients with abandon creates a disaster. We know this biological disaster by the name "cancer."

Cancer cells ignore the normal checkpoints of the cell cycle. Even when essential nutrients run low, cancer cells keep dividing. The result is a tumor.

The body normally produces some cells that will, for various reasons, go from a normal cell into a cancer cell. This process is called transformation. The body's immune system normally spots the abnormal cell and destroys it. Sometimes, however, the cancer cell evades the immune system. Perhaps some cancers grow so fast, the immune system defenses cannot keep pace with their destruction.

The uncontrolled growth of cancer cells has two general outcomes:

➤ Benign tumors, which stay at the original site and often cause no further harm.

➤ Malignant tumors, which continue to grow unabated and spread to other parts of the body (a process called metastasis).

Microbes that experience an out-of-control cell cycle quickly die. But higher plants and animals can get tumors of either type. A person or animal with a malignant tumor is said to have cancer.

Cancer continues to hold puzzles that scientists have not solved. Many unknowns also exist in the cell cycle. The abnormalities of cell growth and division offer almost unlimited questions that need answers.

CHAPTER 13

Genetic Information Travels from Genes to Proteins

In This Chapter

➤ The basics of gene expression

➤ Key points about DNA transcription and translation

➤ What the genetic code means

In this chapter you will learn the critical step in your cells' biology, that is, turning genes into action. Genes carry information, but they seldom roll up their sleeves and get work done in your body. By contrast, proteins seem to do all the work! How do they know what to do and when? The genes tell them.

All your activities result from the information that passes from the genes that make up your DNA to the proteins that operate your body's machinery. Quite a vital undertaking! The main pieces of this information highway belong to two processes known as transcription and translation.

This chapter introduces you to transcription and translation and the way the cell runs them and controls them.

What Does Gene Expression Mean?

Gene expression is the transfer of the information carried in genes to a fully formed protein. Genes are strings of nucleotides that make up sections of the DNA molecule. Proteins are strings of amino acids. How does the cell get from a lineup of nucleotides to a lineup of amino acids? It happens by transcription followed by translation.

An easy way to remember gene expression is to consider this equation:

DNA ➔ RNA ➔ Protein

The "DNA ➔ RNA" step equals transcription. The "RNA ➔ protein" step is translation. Together, they add up to gene expression.

RNA's Job in Transcription

Genes do not build proteins themselves. Genes only spell out a story that enzymes read (transcribe). Other molecules in the cell pass the story on (translate) to the protein-manufacturing part of the cell. Ribosomes play this role as protein factories.

A messenger of some sort seems required to carry the story of life from genes to proteins. In cells, this messenger is ribonucleic acid (RNA). RNA resembles DNA in several ways. RNA consists of a strand of nucleotides, and the specific order of nucleotides represents information or genes. RNA differs from DNA by being composed of only a single strand rather than the helical ladder shape of DNA. In addition, DNA represents your chromosome. DNA contains all the information that defines you as an individual and a member of *Homo sapiens*. RNA simply carries DNA's information during the time your cell is making new proteins.

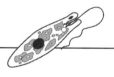

Cells have three types of RNA. The type that reads the information in DNA's genes is called messenger RNA, or mRNA.

The information carried by genes lies in the specific order of the nucleotide bases in DNA, as you learned in Chapter 11. RNA also contains nucleotides bases. In transcription, an enzyme detects each and every base of a gene in order, fishes around to find a complimentary base, and begins building a strand of RNA. In this manner, mRNA gets the information from genes.

RNA's bases connect with complementary bases in DNA in much the same way the two strands of DNA connect with each other. The only difference is that RNA substitutes the base uracil (U) for thymine (T). Thus when an enzyme builds mRNA based on DNA's base sequence, C and G pair the same as in DNA, but A now pairs with U.

With the important genetic message tucked under its arm, mRNA travels to the ribosomes.

Inside a Cell

Three Types of RNA

Cells contain three types of RNA that each have defined jobs in gene expression. Messenger RNA (mRNA) receives information from genes. Transfer RNA (tRNA) selects amino acids to correlate with mRNA's specific bases. Finally, tRNA carries the amino acid to ribosomal RNA (rRNA) in the ribosome. The rRNA acts as the protein assembly site of the cell.

RNA's Job in Translation

Translation converts the information carried in a lineup of mRNA nucleotides to a protein. The key step here involves building a protein of specific amino acids in an exact order. All three types of RNA play a part in this protein assembly, but the most important factor in gene expression relates to the genetic code. This is the way that DNA's composition translates into proteins of their own exact and correct composition.

Why doesn't information flow directly from DNA to protein and eliminate the RNA middleman? It's probably best not to tinker with a molecule as essential to life as DNA. Let RNA poke around the nucleus and take excursions into the cytoplasm to find ribosomes. The DNA should stay safely tucked away in the nucleus, holding an organism's vital statistics as well as the blueprint for the next generation.

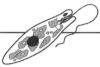

Inside a Cell

Ribosomes

A ribosome is a tiny organelle that has one job: to make proteins. Eukaryotic cells make ribosomes in the nucleolus area of the nucleus. They then ship the ribosomes to the cytoplasm where these little balls cling by the thousands to endoplasmic reticulum. Cells that must secrete large amounts of substances contain tens of thousands of ribosomes. For example, pancreas cells produce digestive enzymes and endocrine glands produce hormones. These cells have thousands of ribosomes.

The Genetic Code

The nucleotoide bases of DNA make up what biologists call the genetic code. This code is nearly universal across species from bacteria to humans. Put another way, the ribosomes of *E. coli* will make the exact same protein as your ribosomes if given the same DNA molecule to transcribe.

The genetic code relies on segments of DNA containing three adjacent bases. These three-base segments are called codons. If a gene represents a sentence of genetic information, a codon represents one word in the sentence.

Each unique base sequence corresponds to the 20 or so amino acids needed to make proteins. For example, the codon GAG stands for the amino acid leucine, CCT means glycine, and GGG equals proline. In transcription, mRNA converts these codes to CUC, GGA, and GGG, respectively.

The genetic code gets transcribed in the correct direction every time due to the DNA-reading enzymes and protein synthesis enzymes. These enzymes always read DNA in one direction. Ribosomes similarly read RNA in only one direction.

The genetic code accepts more than one codon for an amino acid. Biologists call this redundancy. At the same time, codons never stand for more than one amino acid, ensuring that your genetic information will not be confusing. Those same biologists sum up the genetic code as having "redundancy, but no ambiguity."

Biology Vocabulary

Codons

Why did evolution develop a genetic code based on the three-base codon? Why not a one-, two-, or four-base version? Only four nucleotide bases are available, but we need 20 amino acids to make proteins. Clearly, a one-base codon would code for only four amino acids. Two-base codons work better; they would stand for 16 different amino acids. But this still isn't enough. A codon built of three nucleotide bases can stand for 64 possible meanings, more than enough to cover the 20 amino acids.

Because of the three-base codons upon which we base the genetic code, to make a protein of 100 amino acids, we need a segment of DNA 300 nucleotides long.

Other Components of Gene Expression

You might not be surprised to learn that many additional molecules keep transcription and translation working properly. Gene expression is simple in principle but complicated in practice. Some of the additional features of gene expression are:

➤ The enzyme RNA polymerase reads DNA's genetic code and transcribes it to a new strand of mRNA.

➤ As RNA polymerase moves down the DNA, it adds nucleotides to the mRNA, which gets increasingly longer. The lengthening of new mRNA is called elongation.

➤ The enzyme knows where to begin reading on the DNA molecule by finding a section in DNA called the promoter. A similar section makes the enzyme stop reading. This stop sign is called a terminator.

➤ The actual stretch of DNA that the enzyme transcribes is called the transcription unit.

➤ At the ribosome, tRNA reads codons and picks up the correct amino acid corresponding to the codon's base sequence.

➤ An enzyme adds amino acids to the growing *polypeptide* one at a time until a fully formed protein has been built.

Protein: The Final Product

Chapter 2 pointed out that proteins have specific features to make them work. Proteins will not carry out their jobs without a correct primary, secondary, and tertiary structure. You cannot get by without proteins. The information carried in genes goes via RNA to proteins. Genetic information never goes directly into fats or carbohydrates. The body makes proteins first, and then specific proteins are responsible for building fat, carbohydrates, membranes, and new molecules of DNA and RNA. Proteins rule!

A protein's three-dimensional tertiary structure is known as protein folding. The protein must be folded at specific points along its amino acid sequence, and re-folded at additional sites. These folds help a protein recognize its substrate and bind with it. In other words, folding gives the protein its ability to function.

Denaturation refers to the unfolding of a protein to take away its power. The inside of a cell does a good job protecting proteins against denaturation, but if a protein is secreted into nature several things can destroy it:

➤ Heating.

➤ Alcohol, acids or bases, and other chemicals.

➤ Heavy metals such as mercury, lead, and cadmium.

➤ High salt concentration.

➤ Enzymes.

Denaturation means that proteins don't last forever. When they disintegrate or lose their function, the cell makes more by the processes described in this chapter.

For protein synthesis, every cell needs a variety of amino acids. Your diet supplies these amino acids in protein-rich foods, such as meat, dairy products, eggs, and beans. Protein synthesis also requires lots of vitamins. Vitamins help the protein-building enzymes work. In turn, vitamins often need to be activated by certain minerals. To run the protein machinery takes energy, which comes mainly from carbohydrates and fats.

Therefore, good nutrition is the only way to power cell growth its key steps of DNA replication, transcription, and translation. Later chapters emphasize this recurring theme: all the body's processes work together in coordinated fashion to make up your physiology.

CHAPTER 14

Genetics and Inheritance

In This Chapter

➤ The basics of heredity, genetics, and inheritance

➤ Mendelian genetics and how we inherit our traits

➤ How traits and chromosomes relate during meiosis

➤ Linked genes and sex-linked genes

In this chapter we will apply the basics of genetics covered in the previous few chapters to the processes that make you who you are. The genetic code is standard across all of biology, yet no two organisms are alike. How does so much variation fill the biological world when we all begin with the same four nucleotide bases?

Genetics is rooted in your DNA, but inheritance is tied to your chromosomes. In meiosis and mitosis, a lot of chromosome division and separating goes on in our cells. Through it all, your cells manage to keep intact your body's operation manual, its DNA. But variation has been built into the system. This leads to diversity between two members of the same family, among individuals in the same species, and across different species.

The traits you inherit are determined in subtle steps in meiosis and mitosis. You have seen the detailed picture of these processes. Now it's time to consider a bigger picture of inheritance.

Heredity, Genetics, and Inheritance

Heredity is the passage of traits from one generation to the next. Scientists (geneticists) study the features of heredity. Their field of study, called genetics, covers how heredity works and also why it varies from individual to individual.

You cannot control your genetics; it resides in the composition of your DNA. It makes up your genotype. Neither can you control heredity. Your parents determine the traits you will inherit at the moment of fertilization. By the time you have learned enough biology to even think about inheritance, you have already traveled a long road of gene expression!

Biology Vocabulary

Inheritance

Inheritance is the receipt of certain qualities we get from our parents and past generations. Our parents' DNA holds genes for these qualities or traits. Their children then inherit a mixture of both parents' genes. All individuals behave based solely on the genes they inherit and express. If you do not have the gene for blue eyes, you will not have blue eyes as an adult.

Despite having no control whatsoever over their genetic makeup—or perhaps because of this—people are fascinated by the traits they inherit from their parents and ancestors.

Inheritance tells us a bit about where we come from and hints at what future generations will be like. Inheritance also illustrates the ongoing phenomenon of evolution. We forget sometimes that evolution did not stop when humans showed up in history. Evolution continues. Organisms adapt to changing environments based on traits they inherit and other traits that gradually disappear from their population.

Mendelian Genetics

Modern genetics owes its foundations to Gregor Mendel. This nineteenth-century monk investigated how pea plants pass on traits to their next generation. His experiments led to the birth of the science of genetics. Some biologists refer to this field as Mendelian genetics.

Mendelian genetics begins with the proposition that all the traits you inherit are rooted in your chromosomes. How chromosomes replicate and divide in your cells determines the traits you ultimately receive.

Biology Vocabulary

Traits

Traits are any detectable characteristics that an organism inherits.

Mendel discovered several truths of genetics even though he did not have the technology to examine meiosis and mitosis. Now scientists have identified the ways in which these two processes affect chromosomes:

➤ Meiosis—Paired sets of chromosomes separate into single sets; diploid cells become haploid cells. In meiosis, pieces of chromosomes break off and move to the corresponding site on the homologous chromosome. This allows for redistribution of genes into the reproductive cells that will take part in fertilization.

➤ Mitosis—Paired sets of chromosomes replicate and then separate in cell division. Overall, diploid cells produce new diploid cells. This process allows the body to produce new cells for growth and tissue repair.

The variation that we see among all individuals of a species comes from two sources. First, meiosis redistributes genes on chromosomes about to be part of a reproductive cell. Second, these reproductive cells—eggs and sperm cells—get together in a purely random fashion. The randomness of fertilization further adds to the variety of genes you can potentially get from your parents.

Mendel experimented with how genes distribute to offspring by following specific traits in pea plants. These traits were associated with specific genes in the pea plant's DNA. By monitoring easy-to-see traits such as the pea seeds' color and smoothness, Mendel could follow a trait from generation to generation. Since pea plants grow fast, Mendel could run more experiments than if he had chosen elephants as his model!

Mendelian genetics explains why members of the same family all vary a bit in their traits. The explanation comes from two laws of genetics that Mendel's experiments illuminated: the law of segregation and the law of independent assortment.

Biology Vocabulary

Diploid and Haploid

A diploid cell is a cell that contains two sets of chromosomes. Biologists use the shorthand "2n" for the diploid number of chromosomes. In all animals that reproduce by sexual means, the diploid chromosomes come from an organism's parents. One parent supplies one set of chromosomes and the other parent supplies the second set. Together, they form a full diploid set. In humans, this is 46 chromosomes. Almost all cells in your body are diploid. The exception occurs in the reproductive system.

The reproductive system produces haploid cells, called gametes, which contain only one set of chromosomes. Biologists use the shorthand "n" to denote a haploid cell. The female eggs from the ovary and the male spermatozoa from the testes are examples of haploid gametes.

Law of Segregation

Remember that chromosomes come in pairs. The same genes appear on each paired (homologous) chromosome, and each corresponding gene is called an allele. Alleles may be very, very similar to each other, but not necessarily identical.

In meiosis, each allele separates from its mate. This process is called the segregation of alleles. After the alleles have been segregated, the body distributes them into reproductive cells. Therefore, segregation of similar but non-identical alleles leads to variation among all the reproductive cells produced by a male or female.

Biology Tidbits

Gregor Mendel

Gregor Mendel (1822-1884) was an Augustinian monk and science teacher. Mendel was not a trained researcher but rather a lover of nature who took an interest in the variety of flowers that bloomed outside his monastery room each spring. With an almost casual air, Mendel began planting various seeds to see what traits each new generation of plants exhibited. The interest may have turned into an obsession because, focusing on pea plants, Mendel experimented for seven years, recording the characteristics of thousands of progeny from his plant crosses. By tracking the patterns of how traits appear and regress, Mendel singlehandedly invented the science of genetics. At the same time, he laid the foundation for the principles of inheritance.

Law of Independent Assortment

Consider an organism with only two paired chromosomes. Label the chromosomes Yr and Rr. The "Y" and "R" labels and the "y" and "r" labels help you tell apart each homologous chromosome. The "Y" represents a plant that produces yellow seeds and "y" means green seeds. The "R" stands for a plant that makes seeds with a smooth surface and "r" is for a plant with seeds having a wrinkled surface.

Meiosis separates all these homologous chromosomes. The resulting reproductive cells (gametes) then get one half of the original set of chromosomes at the end of meiosis. This sorting happens independently, meaning the chromosomes do not stick to each other or travel together. Like a shuffled decks of cards, you have an equal chance of getting any combination of alleles from your parents. In our example organism with two sets of chromosomes, meiosis produces eggs or sperm cells with the following alleles in just about equal distribution:

➤ 25 Percent are YR.

➤ 25 Percent are Yr.

➤ 25 Percent are yR.

➤ 25 Percent are yr.

Fertilization recombines the alleles at random to make a new generation of offspring.

What did Mendel observe when he crossed two different pea plants to produce a new generation of plants?

Following Traits through Generations

Mendel began with two different pea plants that he distinguished by the appearance of their seeds:

➤ Plants containing a YYRR diploid set of chromosomes produced yellow seeds with smooth surface.

➤ Plants containing a yyrr diploid set of chromosomes produced green seeds with wrinkled surface.

Each plant went through meiosis and produced gametes. The yellow/smooth-seeded plant made gametes with a chromosome like this: YR. The green/wrinkled-seeded plant's gametes looked like this: yr. Mendel crossed these two plants, meaning one fertilized the other.

When the next generation of plants underwent meiosis, a new set of gametes appeared. These gametes possessed the breakdown of alleles shown above; each had a 25 percent chance of receiving YR, Yr, yR, or yr.

To figure out which seeds received which traits, Mendel crossed this generation of offspring. The random fertilization produced the following results:

➤ Nine plants had yellow and smooth seeds.

➤ Three plants had green and smooth seeds.

➤ Three plants had yellow and wrinkled seeds.

➤ One plant had green and wrinkled seeds.

What did Mendel discover? He saw that some traits (yellow and smooth) seemed to dominate other traits (green and wrinkled). Rather than producing a generation of yellowish green and partially wrinkled seeds, he made a generation that inherited clearly defined traits. We now understand a basic piece of information about inheritance: some traits dominate other traits after fertilization has taken place and an embryo begins to develop.

Biology Tidbits

Genetic Testing

Scientists have improved their ability to identify and predict health disorders by examining chromosomes. This is called genetic testing. In genetic testing, a scientist looks at two main features of the chromosomes: abnormal chromosome number and unusual chromosome shape. For example, Down syndrome results from a person having an extra chromosome 21, so that the body's cells contain 47 total chromosomes instead of 46. This disorder is an example of aneuploidy, an abnormal number of a particular chromosome. A general term for any abnormal number of chromosomes is polyploidy.

Recessive and Dominant Traits

Biologists have followed in Mendel's footsteps, crossing peas, fruit flies, grasshoppers, cockroaches, and every other darn thing they could catch and carry back to a lab. These Mendelian experiments have shown that many traits can be classified as either recessive or dominant.

A recessive trait is a characteristic (called a phenotype) that does not normally show up in a population. A dominant trait is a phenotype that shows up frequently in a population. For example, in people, brown eyes are dominant over blue eyes.

But wait a minute! Plenty of people have blue eyes. How did all those recessive traits appear? Why haven't all the brown-eyed people outnumbered the blue-eyed ones?

In Mendel's first generation of pea plants, YR and yr gametes combined. All the offsprings' chromosomes were YrYr. Yellow (Y) is a dominant trait over green (y). You need only one dominant allele in a homologous pair to get yellow seeds. Smooth (R) is dominant over wrinkled (r), and only one R allele is needed to make smooth seeds. To produce offspring in which recessive traits appear, the offspring must get two recessive alleles. In our example, only two y alleles together produce a green seed and only two r alleles make a wrinkled seed.

Because of the law of independent assortment, the next pea generation came from gametes that each had a 25 percent chance of being YR, Yr, yR, or yr. Mendel first cross-produced eight gametes with this distribution of alleles.

When you complete all the possible combinations of the eight gametes, you end up with 16 possible combinations of alleles. Here are the allele combinations that Mendel produced:

➤ YYRR = yellow-smooth.

➤ YyRr = yellow-smooth.

➤ YYRr = yellow-smooth.

➤ YyRR = yellow-smooth.

➤ YyRr = yellow-smooth.

➤ yyrr = green-wrinkled.

➤ Yyrr = yellow-wrinkled.

➤ yyRr = green-smooth.

➤ YYRr = yellow-smooth.

➤ Yyrr = yellow-wrinkled.

➤ YYrr = yellow-wrinkled.

➤ YyRr = yellow-smooth.

➤ YyRR = yellow-smooth.

➤ yyRr = green-smooth.

➤ YyRr = yellow-smooth.

➤ yyRR = green-smooth.

Imagine the variety of traits a person can have when compared with this example of only two traits in peas!

How Genes Are Linked on Chromosomes

After I have made your head spin with laws of segregation and independence, here's another nugget to ponder. Geneticists now realize that some genes located on the same chromosome tend to be inherited in a package deal. These are called linked genes and they seem to contradict Mendel's law of independent assortment.

Experiments with fruit flies—their scientific name is *Drosophilia*—have uncovered numerous linked genes. These traits relate to the flies' eye color, wing size, body color, and so on.

Geneticists have solved a large portion of the puzzle regarding which genes link to others by making maps of entire chromosomes. A gene map shows where a specific gene sits on a chromosome. The nearness of two genes increases the chance that the two genes are linked. As a result, a newborn either has both traits or neither. Genes that are situated far apart on a chromosome or on different chromosomes have less chance of being linked.

Biology Tidbits

Chromosome Gene Maps

Geneticists have produced detailed maps of all 23 chromosomes of humans and quite a few sets from other species. Many of these maps show the region of a given chromosome where a gene for a disease occurs. All you needed to do is find an online resource for "chromosome maps of genes" and you can view diagrams of your genetic makeup!

Sex-Linked Genes

Plenty of jokes exist on the subject of sex-linked genes. These are genes that are located on a person's sex chromosome. Let's take a closer look at the sex chromosomes.

Being male or female is one of the more obvious phenotypes in humans. This trait is determined by a single chromosome. In humans and other mammals, an individual carries either an X chromosome or a Y chromosome. Thus a human has 22 homologous chromosomes plus a pair of sex chromosomes.

In the testes and the ovaries, the two sex chromosomes segregate during meiosis. Each gamete receives one of these chromosomes. Half of all sperm cells carry an X chromosome and the other half carry a Y. All eggs carry the X chromosome. Thus the type of sperm cell (X or Y) that fertilizes an egg determines the infant's gender.

A person who has inherited two X chromosomes from their parents becomes a female. A person with one X and one Y chromosome becomes a male. (Recently, geneticists have discovered some exceptions to these rules.)

Meanwhile, other genes occur on the X and Y chromosomes. Some of these genes are expressed only in offspring of a certain sex. So, let's reconsider the difference between linked genes and sex-linked genes. Linked genes are genes on the same chromosome that tend to be inherited together by either male or female offspring. Sex-linked genes are genes found only on the sex chromosomes and their associated traits may be more likely or less likely to appear depending on your sex.

Geneticists discover new sex-linked genes as they do research on the human chromosome. Three well-studied examples of sex-linked genes are:

➤ Color blindness—Linked to the X chromosome as a recessive trait, meaning it is more prevalent in males and a male can only be colorblind if his mother carried the gene for it.

➤ Hemophilia—Also an X chromosome-linked recessive trait. In this disorder, a person lacks certain proteins in the blood necessary for forming clots.

➤ Duchenne muscular dystrophy—This X chromosome-linked trait results in a progressive weakening of the muscles due to the absence of a muscle protein called dystrophin.

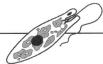

Inside a Cell

Sex Chromosomes

In a high-powered microscope, the sex chromosomes differ markedly in size and shape. The X chromosome looks like other chromosomes belonging to the mammal. The Y chromosome by contrast is much smaller. Both sex chromosomes in humans have been mapped so that scientists know the location of all genes on them. Although the Y chromosome is short, it contains regions homologous with the X chromosome. This allows it to behave just like other chromosomes during meiosis.

Some sex-linked genes occur on the Y chromosome. Most of these genes are, however, associated with sterility and therefore are not passed to a new generation.

PART FOUR

Adaptation and Evolution

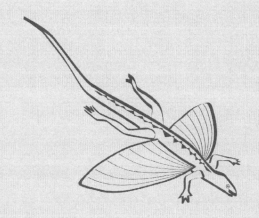

CHAPTER 15

Mutation and Other Genetic Errors

In This Chapter

➤ A review of mutation and different types of mutation

➤ The meaning of adaptation and its connection to mutation

➤ An introduction to natural selection

In this chapter you will see why mutation is a normal and necessary part of evolution. The thought of mutants can scare people, especially since Hollywood never fails to invent new mutants to terrorize mankind. Mutations nevertheless can make species stronger and more fit for surviving in their environment.

Begin by learning what mutation means to biology. By understanding mutation, you will see how it relates to evolution and why it gives meaning to the phrase, "survival of the fittest."

What Is Mutation?

A mutation is any error in the genetic code. By this, I mean a mutation can be any incorrect sequence of nucleotides in DNA so that a gene's meaning gets lost.

Think of the following sentence as a gene composed of words (codons) that are each made up of specific letters (nucleotide bases: "The fat cat sat on the hat.").

If you were to take a word or two out of the sentence, the gene might still retain its meaning but it would differ from the original sentence. For instance, "The cat sat on hat," gets the job done. But it does not communicate the message as clearly as the correct sentence, does it?

This is what some mutations do; they alter a gene without removing all of the gene's main information.

A more serious mutation might do the following to the sentence: "The sat on the." Now your body is totally confused about what this gene is trying to say to the ribosomes. Of course, the gene might disappear from your DNA altogether, which would be the most critical type of mutation.

Scientists have identified certain consistent patterns in mutations. These types of mutations range from minor to very serious distortions of a gene's message.

Types of Mutations

Mutations occur in segments of a gene or they involve entire genes. In either case, a mutation relates to an incorrect nucleotide sequence in DNA:

➤ Substitution: "The fcg cat tcg on the haa." In this example, the wrong letters (nucleotide bases) have been substituted for the correct ones.

➤ Insertion: "The fattttt cat sataagc on the hat." Extra bases have been inserted where they do not belong.

➤ Deletion: "The f ca sat on th at." Some bases are missing.

➤ Frameshift: "hef atc ats ato nth eha t." The normal arrangement of three-base codons has shifted and the gene's meaning disappears.

Living things would not have made it this far without some resilience. Cells have mechanisms to extract the meaning from damaged genes, but the examples above show that the situation is sometimes irreparable.

What Causes Mutations?

Some agents are known to cause these types of mutations in DNA. For example, ultraviolet light, radioactive radiation, some chemicals, and some viruses can cause mutations directly in the DNA molecule.

Other mutations happen when the enzyme involved in DNA replication makes a mistake. The enzyme makes a new copy of the DNA with the mistake as part of its molecule. All subsequent DNA replications might retain the error and generations of offspring retain it also. If the mutation has a large effect on the offspring's physiology, it could lead to a new species. This event is called speciation and is quite common in microbes.

Biology Vocabulary

Speciation

Speciation is the origin of a new species in evolution. It occurs often in microbes that tumble through a dozen generations in the time it takes a person to fly across the United States. Although speciation may appear rarer in higher organisms, quite a bit occurred during evolution. Scientists estimate the Earth has about nine million species of plants and animals. This could be an underestimation, however, because many species remain undiscovered. In addition, while many species go extinct every day, others branch off from existing species. The exact number of species on Earth may never be known.

Mutation Rate

The speed with which a species changes its traits relates to mutation rate. Mutation rate equals the number of mutations per 100,000 genes every time a new generation is born.

Prokaryotes and viruses have the highest mutation rates. These organisms have short generation times, so mutations can quickly give rise to new species. In fact, new species of bacteria come and go much faster than in any other organism.

Multicellular eukaryotes have the slowest mutation rates. In animals and plants, mutations occur about once every 100,000 genes per generation.

How Mutation Leads to Adaptation

An animal or a plant has available to it only the genes its population contains. This total amount of genes in a population is called a gene pool. What happens if an animal could benefit from a better gene, say, one that improves eyesight? Perhaps this animal normally eats grasshoppers but all the grasshoppers in its habitat have been wiped out by pesticides. The gene pool contains no genes for improved eyesight, so our unlucky animal cannot see the aphids all around it! In this scenario, the myopic animal could well go extinct.

Mutations usually happen by chance and not through planning. (Only certain bacteria have the ability to affect their own mutation. They do this by speeding up their growth rate, which in turn speeds up the mutation rate. The bacteria keep this up until they hit upon a mutation that helps them survive.) Imagine the impact of a chance mutation in the eyesight gene of one grasshopper-deprived animal. This animal can now fatten up on aphids while his friends go hungry. The ability to see aphids is an adaptation. In this admittedly improbable example, the adaptation means the difference between life and death. In a larger sense, adaptation might mean the difference between species survival and extinction.

Biologists cite hundreds of examples of plants and animals that have adapted to conditions of an environment changed by humans. Species that have the ability to adapt have a better chance at survival; species unable or slow to adapt creep closer to extinction.

Biology Tidbits

Adaptation

The species that adapts to changes in its environment has an advantage over species that cannot adapt. This concept has proved especially true since the Industrial Revolution, when human activities began to drastically change the environment.

One of many examples of adaptation comes from a comparison of the burrowing owl and the American crow. The owl lives on the ground and uses den openings vacated by ground squirrels. Its habitat is undisturbed grasslands. For food, it seeks mainly the insects and small reptiles of grasslands. Housing developments, roads, and other features of urban sprawl have destroyed much of the owl's range. As a result, burrowing owl populations are in a precarious situation. By contrast, the American crow thrives in urban areas. It picks through garbage and landfills with glee and hardly seems perturbed by humanity's commotion just a few feet away. Biologists call the owl a specialist and the crow a generalist.

Because of the genetic makeup of crows that allows them to be generalists, they also have the capacity to adapt. By adapting both their physiology and their behavior, crows will undoubtedly survive long into the future. I cannot make the same prediction for the burrowing owl.

Biology Vocabulary

Genetic Variation

Mutation increases genetic variation. By this I mean that the size of the gene pool increases and each generation exhibits more variation among individuals than a population with a limited gene pool. To see genetic variation, you need look no farther than at dogs. This variation enables the dog species (*Canis lupus*) to be extremely adaptable.

An Introduction to Natural Selection

Natural selection is the ability of certain members of a species to do better in the environment than other members of the same species. Nature "selects" them to live longer and produce more offspring. In time, this natural selection improves the overall survival of the entire species.

The nineteenth-century biologist Charles Darwin emphasized the effect of natural selection on reproduction. By his view, the most successful individuals also possessed heritable traits that they could pass on to their offspring. In fact, natural selection does not mean much to a species unless the helpful traits can be passed along to future generations.

From a geneticist's view, natural selection occurs because certain favorable alleles show up in a gene pool at a higher frequency than other alleles. As the gene pool evolves, so does a species. Evolution would not have occurred without natural selection. At the same time, a species' success rests on its adaptations, and adaptations come from mutation.

CHAPTER 16

Evolution and Extinction

In This Chapter

➤ How the theories of evolution developed

➤ Main points of Darwin's theory of evolution

➤ How the origin of species concept works

➤ The meaning of extinction and types of extinction

In this chapter this book reviews the broad subject of evolution. I stick to the highlights and main concepts of this complex topic. This chapter also covers extinction. Why talk about extinction with evolution? Because they work in tandem.

Evolution is an ongoing process. It led to where biology is today and it will continue past human history. Throughout all of evolution occurs the development of new species and the extinction of others.

Even the Theory of Evolution Evolved!

Science operates on theories proved or disproved by experimentation. Not all experiments work as expected. Some give the results expected, but more often experiments produce disappointing results or an outcome that truly astounds. Scientists build their theories on the accumulated data from hundreds and sometimes thousands of studies.

Biology Tidbits

Charles Darwin (1809-1882)

Charles Darwin's famous theory took shape beginning in 1831 when he accepted a job to board the *HMS Beagle* and chart the coastline of South America. But Darwin looked past the coastline. He recorded the numerous plant and animal species of the tropics and compared them to fossils he found along the way. In his spare minutes, Darwin recorded his observations on the geologic features of the lands the *Beagle* visited.

Darwin concluded that the Earth could not possibly be as young as theologians said. To develop the diversity of life he saw, Darwin supposed life took eons to reach its present state. Darwin insightfully surmised that geology, geography, and biology influenced each other.

Darwin was the first biologist to focus on the importance of adaptation to the development of species. By the 1840s, he had completed most of his theory on natural selection but he hesitated to publish it. Perhaps Darwin sensed a comprehensive theory linking the evolution of life to the evolution of the planet would be a big pill to swallow for the times.

With some prompting from his contemporary Charles Lyell, Darwin published *The Origin of Species* in 1859. The first printing of 1,250 copies sold out in one day. It caused an uproar as he had predicted. The book also opened science to the concept of *biodiversity*. Darwin's theory of evolution remains the central theory that explains biology on Earth.

The theory of evolution evolved before scientists had all the technology they needed to demonstrate their ideas. In addition, science before the seventeenth century included a significant amount of religious theory. As the theory of evolution has evolved over the past 200 years, scientists have struggled to accommodate religious beliefs. Religion influences science less than in the past, but scientists still face the challenge of sorting through massive amounts of data to find scientific truth.

If you are confused by evolution, you are not alone. Evolution includes large amounts of fact supported by data. But where gaps in our knowledge exist, theory fills the gaps. This is true for all areas of science, including medicine. Future generations of scientists will run experiment after experiment until they fill in all the gaps, if possible, with data-based facts.

Catastrophism

In the eighteenth century, French scientist Georges Cuvier developed the theory of catastrophism to explain how life formed on Earth. At the same time he laid the foundation for paleontology, the study of fossils.

Cuvier dug through rock layers in the vicinity of his home in Paris. As he dug deeper, he found fossils, noting that the deeper the rock layer, the more dissimilar the fossils appeared from modern forms of life. Cuvier proposed that distinct layers of rock each represented a period in Earth history. Each layer, or time period, was furthermore defined by some natural catastrophe that wiped out most of the life that came before it.

By catastrophism, evolution has proceeded in fits and starts, interrupted by cataclysmic disasters. Good material for a motion picture, but catastrophism soon gave way to other theories that included more gradual progression of life on Earth.

Gradualism

Proponents of gradualism took an opposite approach to evolution from catastrophism. This theory was proposed at the close of the eighteenth century by Scottish geologist James Hutton. Hutton proposed that the current Earth had resulted from gradual, inexorable changes over time. For example, a valley could form after centuries of a river's flow washing through it. Hutton surmised that life also developed as a gradual process.

About 50 years later, geologist Charles Lyell built on Hutton's concept. Lyell proposed that the Earth's geologic changes have always been fairly uniform. The geology today is a progression from the geology of earliest Earth. Lyell's theory of uniformitarianism stated that today's Earth is changing now at the same rate as it has always changed since its birth.

British biologist and theologian Charles Darwin took note of Lyell's theory. Like Lyell, he believed gradualism could mean only one thing: the Earth must be much older than 6,000 years, the period proposed by theologians who estimated Earth's age by adding up the generations described in the Bible. All sciences entered a period of excitement and debate in the 1800s. Theories on evolution were proposed, argued; some were rejected and some were accepted. Science and theology took separate paths, never to again intersect.

Lamarck's Ideas on Evolution

French biologist Jean-Baptiste de Lamarck raised eyebrows by publishing a new theory in 1809. Lamarck had compared fossils of ancient life with modern life. The differences he saw led him to make two astounding proposals:

> ➤ Life evolves through use and disuse of body parts; useful parts stay and little-used parts wasted away and disappeared from future generations.

> ➤ Life has a tendency to become more complex during evolution. Thus, organisms inherited traits that added complexity and rejected traits that simplified their body.

The next generation of scientists would discount Lamarck's theories; some did so in rather cruel fashion. Lamarck was nevertheless on the right track. His use-disuse theory resembles the idea of adaptation and natural selection, soon to be proposed by Darwin. Lamarck's ideas on complexity agree with the evolution of gradual discrete changes over a long period of time.

Darwin's Theory of Evolution

Biologists have refined the theory that Darwin developed over more than 20 years of recordkeeping. These central themes of this theory of evolution remain, however, unchanged:

➤ Life comes from life.

➤ Populations of organisms adapt to changes in their environment through evolution.

➤ Evolution involves a change in the population's genetic makeup.

➤ The genetic makeup changes through successive generations.

➤ All species evolved—biologists say they "descended"—from an ancestor species.

Biology Tidbits

What We Learn from Fossils

Fossils are petrified skeletons, bones, feathers, teeth, shells, seeds, leaves, or rock impressions of these things from organisms that lived long ago. Fossils offer evidence of ancient life and give scientists an idea of the organism's physical structure. Unfortunately, fossils are rare. Even the fossils found by paleontologists are usually missing portions of the plant or animal body.

Scientists now examine the chemical makeup of fossils and try to recover any DNA samples associated with the fossil. By doing these detailed studies, biology builds a more complete picture of the organisms that lived in Earth's early history.

Remember also that populations, not individuals, evolve by changing their genetic makeup.

The Origin of Species

Darwin used the word "evolution." He described the development of life on Earth as "descent with modification." This simple phrase goes to the core of evolution, that all present-day organisms developed over time from a common ancestor. We now call the modifications their adaptations.

You may have heard of the evolutionary tree of life. Picture a multi-branched tree turned upside down. The common ancestor sits at the top. The main trunk descends from it and as it does, it forks and branches extend outward from each fork. The forks stand in for main routes of evolution where different types of organisms separated from each other during their development. The main branches have several smaller branches. Each main branch represents a group of related organisms. The small branches represent closely related organisms.

A century earlier, the Swedish physician Carolus Linnaeus had developed a similar scheme for organizing all the world's life. Darwin used the same rationale that Linnaeus used when he proposed descent with modification. The modifications produced new species. Natural selection determined the species that would succeed and those that would fail or go extinct.

The tree of life illustrates a concept called macroevolution, which relates to how all species

Biology Vocabulary

Macroevolution and Microevolution

Evolution can be thought of on a global scale and at the level of a single species' population. Biologists use the terms macroevolution and microevolution for these ideas, respectively. Macroevolution refers to long-term, large-scale changes in many species. Microevolution refers to small changes in a single population's genetic makeup.

Biology Tidbits

Carbon Dating

Scientists determine the age of fossils by a technique called carbon dating. The element carbon contains six protons and six neutrons that together give a carbon atom a weight of 12. But nature also has a small percent of carbon with extra neutrons. This is called carbon-14. Scientists know carbon-14 takes 5,730 years to convert back into regular carbon-12. Scientists measure the ratio of carbon-14 to carbon-12 in a fossil and plug the value into an equation. The answer gives the age in years of the fossil.

develop over time. By contrast, small changes that add variability to a population belong to microevolution.

The Relationship between Evolution and Biodiversity

Adaptations make evolution go forward. Adaptation gives a population an advantage over others of the same species. The adapted population therefore succeeds in the environment where others may fail. Eventually, the adapted population becomes the only population in a particular environment.

Some adaptations give a population an extraordinary advantage. The adaptation might enable the population to live in a different habitat, eat a different diet, or behave in a unique way. If an adapted organism is different enough from other members of the species, the adapted individual might start a new species.

Several types of adaptations have led to new species, and new species increase overall biodiversity:

> ➤ Structural adaptation—A change in the outward appearance of an organism, such as unique pigmentation or the growth of a shell.

> ➤ Physiological adaptation—A change in metabolism, such as the ability to hibernate over the winter.

> ➤ Behavioral adaptation—A change in an organism's behavior, such as the urge to migrate.

As adaptations make species look, function, or behave differently, biodiversity increases.

Biology Vocabulary

Fitness

You might think of fitness as a goal to achieve by going to a gym. In evolution, fitness has a different meaning. Fitness is the contributions an individual makes to a gene pool that will benefit future generations. Think of the phrase, "He has good genes." This is a way of saying a person has fitness. If this wonderful person has children, his "good genes" help improve the human species.

Extinction

Extinction is a normal part of evolution due to natural selection. Just as the fittest species survive changes in their environment, species that struggle to withstand the changes might not make it into the future. Put another way, an individual with a high degree of fitness has a good chance of survival; an individual with poor fitness has a poorer chance of survival.

Types of Extinction

Extinctions occur in different ways. All of these types have taken place many times in Earth's history. In fact, biologists estimate that 99.9 percent of all the species that ever existed are now extinct. Extinction is part of natural macroevolution.

Which of the following types of extinction do you think humans will experience someday?

> ➤ Background extinction—A small number of species disappear continually at a constant and low rate, about one species for every million species on Earth. This extinction is due to local environmental changes to which the species cannot adapt.

> ➤ Mass extinction—A large number of species disappear at an elevated rate above background rates. A widespread or global natural disaster can cause this extinction. Mass extinctions can remove more than 50 percent of existing species over a span of 10 million years. In Earth's history, there have been five mass extinctions.

> ➤ Mass depletion—A large-scale disappearance of species similar to mass extinction. Depletions do not reach mass extinction levels because they are either not as widespread or they do not involve as many species.

Biology Tidbits

Endangered Species

An endangered species is a species with so few members in the wild, it will likely go extinct in a region of the world or everywhere. Some species are called "extinct in the wild," meaning they have already disappeared from their natural habitats and live only in zoos or other sanctuaries.

Extinction is a natural process in evolution. Unfortunately, humans have altered the patterns and rates of extinction. Biodiversity expert Edward O. Wilson pointed out, "The first animal species to go are the big, the slow, the tasty, and those with valuable parts such as tusks and skins." Biologists have tried to reverse the extinction crisis by enacting laws such as the U.S. Endangered Species Act. But laws that try to control nature can be flawed, and no single law or group of laws can undo the constant threat to biodiversity posed by humanity.

Speciation is the opposite of extinction. In speciation, biology creates new species. As long as the rate of speciation stays ahead of the rate of extinction for all species on Earth, life here will continue. But some species cannot keep up with the rate at which populations are declining. We know these as endangered species.

CHAPTER 17

How We Organize Species

In This Chapter

➤ Learning about how living things are classified

➤ The categories biologists use to organize living things

In this chapter you will learn what to make of all the millions of species everywhere around you. How can anyone possibly keep them all organized so that we understand how humans are like chimpanzees but unlike tortoises?

Carolus Linnaeus developed the first logical system for organizing all living things based on how they relate to each other. He did this without the benefits of DNA analysis. Today, scientists analyze the composition of DNA, RNA, mitochondria, and ribosomes to refine Linnaeus' original classification scheme. We still, however, continue to use the basic framework of classification that Linnaeus built.

Phylogeny, Systematics, and Taxonomy! Oh My!

These three fields of science all relate to the classification of living things. Subtle differences exist between each of these disciplines that perhaps only a scientist could appreciate. I will give you the highlights of each:

➤ Phylogeny—The evolutionary history of a species.

➤ Systematics—The science of studying the relationships of present-day organisms and their relationship to extinct organisms.

➤ Taxonomy—The science of assigning organisms to categories based on similar and different characteristics, and leading to a classification scheme. Taxonomy also includes the conventions used for naming organisms.

Biology Vocabulary

Species

A species (plural is also "species") is the most basic unit in the classification of organisms. A species is a group whose members have similar characteristics and have the ability to breed with each other.

To be a member of a species, an individual must resemble other members of the species. Today's sophisticated analysis methods use genetics more than appearance to classify many organisms. Appearance nevertheless is still used. Organisms such as fungi continue to be classified mainly by appearance.

Every organism that scientists know about has a unique name. Linnaeus devised the naming system biologists use today. They continue to use this system because it not only assigns a unique name to an organism—no two organisms on Earth can have the same name—but it also fits each organism into the hierarchy of living things. An organism's name by the Linnaeus method tells us about the organism's closest relatives and more distant relatives.

Biology Vocabulary

Hierarchy

A hierarchy is a way of classifying things from the top main category to the bottom where several sub-categories exist. Objects at the top might be the most powerful, such as seen in a corporation organization chart. The president sits at the top of the hierarchy. Beneath the president are vice presidents. High- and mid-level managers follow, and then department heads. The hourly staff fill out the bottom of the hierarchy.

How Hierarchies Work

In biology, a hierarchy is an organization scheme for microbes, plants, and animals. Each hierarchy begins with the broadest classification to which all these organisms belong. These all-encompassing categories are called domains. Hierarchies then branch into sub-categories and smaller sub-sub-categories until all organisms have been put into a category. Hierarchies are organized so that similar organisms are in categories situated near each other. As organisms become less related, their categories also move farther apart, like an upside-down tree.

The classification hierarchy from the broadest categories to the narrowest categories are:

Biology Tidbits

How to Name a Species

Carolus Linnaeus invented the binomial method for naming species. By this scheme, every species receives a unique two-part Latin name, for example, *Escherichia coli*. The name of the genus always comes first and is capitalized: *Escherichia*. The second name (*coli*) identifies the species. A genus can contain more than one species, but each species is a unique group of individuals in biology.

➤ Domain

➤ Kingdom

➤ Phylum

➤ Class

➤ Order

➤ Family

➤ Genus

➤ Species

You can take a bottom-to-top viewpoint of taxonomy, too. For example, a species belongs to a genus; genera (plural) are grouped into families; families are grouped into classes, and so on. Each of these levels is called a taxon. For example, an order is a taxon, and so is a genus, as well as a kingdom.

Domains

In the hierarchy of living things, domains are the broadest categories you can use to classify an organism. Biologists have used different numbers of domains in the past 200 years as they learned more about the relationships between organisms. Today, we use three domains:

➤ Archaea—Contains the prokaryotes called archaea.

➤ Bacteria—Contains the prokaryotes called bacteria.

➤ Eukarya—Contains all other types of cells: algae, protozoa, fungi, plants, and animals.

Kingdoms

Kingdoms comprise the second major classification branch in the living world. Biologists of the past century to the present have devised several different ways of putting life into kingdoms. Even today, books and Internet resources might differ on the kingdoms they describe. Some resources refer to five kingdoms and others use four. Yet another school of thought uses kingdoms and domains interchangeably. In this last method, the kingdoms contain prokaryotes, animals, and plants.

The debate on which life forms should go into kingdoms continues. A common scheme you might see in many biology books uses the five-kingdom system:

➤ Monera—The prokaryotes, which have simple, single cells with no internal membranes and which ingest their food by absorbing it or use photosynthesis.

➤ Protista—These are single eukaryotic cells, such as an amoeba, that ingest or absorb food or use photosynthesis.

➤ Fungi—Multicellular eukaryotic organisms that can grow filaments and which absorb their food. This includes mushrooms, molds, yeasts, and smuts.

➤ Plantae—Multicellular eukaryotic organisms with the ability to make their own food by photosynthesis but lacking locomotion. This kingdom comprises the green plants, including mosses, ferns, vascular plants, and trees.

➤ Animalia—Multicellular eukaryotic organisms with their own locomotion—some exceptions exist—and must ingest their food. This kingdom covers animal life from sponges to mammals, including humans.

Some biologists prefer to break up Kingdom Protista into photosynthetic and non-photosynthetic members. The photosynthetic protists then get lumped in with plants and the non-photosynthetic protists go with animals.

A Taxonomist's Life

Many scientists have tried their hand at classifying the world's living things. They soon learned that taxonomy can be difficult and maddening. Biology has accumulated lots of information about some species but very little data on most other species. Some species are hard to find, impossible to catch, or deadly. Much guesswork goes into taxonomy to fill the gaps in knowledge about species we cannot study well or cannot reach.

Current taxonomic schemes use a blend of hard science (DNA analysis, cell chemistry, etc.) and intuition. If you were to visit a taxonomist's lab and listen quietly, you might hear many sentences begin with, "It looks a lot like a…"

Taxonomy is furthered complicated by strange occurrences that happen in species. Some species interbreed. At some point, this causes two species to become one. Conversely, a species that has one population on Continent A and a second population on Continent B might never have a chance to mingle and breed. Certain adaptations in one of the populations might make it quite distinct from the population living on the other continent. Over time, the species diverge and taxonomists classify them as two distinct species.

The classification of living things illustrates one area of science where the science is seasoned with a good deal of art.

Inside a Cell

DNA Homology

Biologists estimate the closeness of a relationship between two individuals using DNA homology. In this method, the biologist gets a DNA sample from both individuals. By heating the DNA, the molecule separates into two strands. The scientist can then mix together the single strands from both samples and give them a few minutes to reconnect. Closely related individuals will have very similar DNA and their strands will connect (called annealing) along large segments of the DNA molecule. This is called a high degree of homology. Unrelated individuals will have many differences in their respective DNA molecules. Their DNA strands will not connect well. This is called distant homology.

Easy laboratory methods have been designed to test the DNA homology between any two samples of DNA.

How Humans Fit into Taxonomy

Humans fit into a taxonomic scheme just the same as any microbe, plant, or other animal. When you view your taxonomy, you gain a better appreciation of how all living things relate to each other.

The scheme for humans from the broadest classification of domain to the most specific classification in species is:

> ➤ Domain Eukarya—One of only three domains, this domain contains all organisms made of cells with a nucleus and membrane-bound organelles.

> ➤ Kingdom Animalia—One of five kingdoms, this one includes all multicellular organisms that ingest their food, cannot do photosynthesis, and usually have a means of locomotion.

➤ Phylum Chordata—All members have a central column that houses a nerve cord running from the top (anterior) to the bottom (posterior) of the animal. Members also have a pharynx that keeps food from going into the respiratory tract and a tail that extends past the posterior opening (anus). In other animals, the anus is the most posterior part of the body. (Humans lose their tail during the embryo's development in the womb.)

➤ Subphylum Vertebrata—Members have a brain enclosed in a skull, separate units (vertebrae) that form the nerve cord, a skeleton on the inside of the body (endoskeleton), a closed circulatory system made of vessels and a heart, excretion of metabolic wastes by kidneys, and separate sexes for reproduction.

➤ Class Mammalia—Members are characterized by fur or hair and mammary glands for milk secretion for the young. Mammals also have a *diaphragm*, are warm-blooded, and their young are born live.

➤ Subclass Theria—Members have separate openings for excretion and reproduction.

➤ Infraclass Eutheria—Mothers carry the unborn fetus in a uterus, in which the fetus receives nutrients and oxygen via a placenta and umbilical cord. This infraclass includes primates, cats, dogs, bears, hoofed animals, rodents, bats, and marine mammals.

➤ Order Primates—Features include opposable thumbs and eyes that face forward. This order includes apes, lemurs, and monkeys.

➤ Suborder Anthropoidea—Monkeys make up more than 90 percent of the species in this order. The rest are apes and humans. This category contains the simian primates, meaning the stereoscopic eyes are well-developed, see colors, and the animal depends primarily on sight for sensing the environment. These animals also have sockets that protect the eyes and a short snout. The latter feature gives evidence that these animals rely less on sensing smells than other animals.

➤ Superfamily Hominoidea—This group contains the "great apes," including apes, chimpanzees, gorillas, orangutans, gibbons, and humans. They walk in an upright posture, lack tails, and have a large brain relative to the rest of the body.

➤ Family Hominidae—Includes the chimpanzees, gorillas, and orangutans. These are omnivores that possess a similar arrangement of internal organs and similar blood composition. (Omnivores are animals with a diet of both plant and animal origin.)

➤ Genus *Homo*—This genus refers to "man," which has a relatively larger brain than the other great apes.

➤ Species *sapiens*—The species *Homo sapiens* has a developed language and uses tools. Species of all types are always identified by both the species name and the genus name. This method of naming all species on Earth is called binomial nomenclature.

Notice that genus and species are always written in italics.

The taxonomic scheme for humans contains some categories missing from the general scheme shown earlier in this chapter. Infraclasses, suborders, and superfamilies help taxonomists distinguish among groups of related life that contain hundreds of species.

Classification schemes change often. Bacteria seem to be classified and reclassified almost daily! Even the classifications for humans have been changed as scientists learn more about the path we took in our evolution and the animals most closely related to us. The advances in genetic testing have helped scientists make more accurate estimates of how living things relate to each other. Unfortunately, when scientists reclassify organisms some people interpret this as confusion among scientists. Science is a fluid thing. As technology gives scientists better ways to analyze the universe, these scientists correct errors and add to our ever-expanding knowledge.

PART FIVE

Prokaryotes, Eukaryotes and Viruses

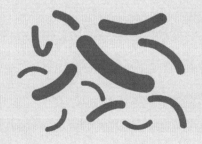

CHAPTER 18

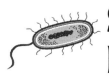 **Single-Celled Organisms: Bacteria and Archaea**

In This Chapter

➤ The main features of archaea and bacteria

➤ Important members of the archaea and bacteria and their features

In this chapter you meet the prokaryotes. Bacteria and archaea are the simplest independently living things. In Chapter 3 you took a tour of a prokaryotic cell and learned about the internal structures that keep these cells alive. This chapter will introduce you to the major groups of bacteria and archaea, and why they are important.

Biology books use one of two ways to introduce readers to prokaryotes. The first method guides you through the taxonomy of prokaryotes. By this method, you would learn about each phylum and the orders, classes, families, and genera in each. A second way to gain insight about prokaryotes is to take them as groups of organisms that perform certain tasks on Earth. I will introduce prokaryotes to you by this second method. This way, you learn how the prokaryotes affect your life and health and shape your environment.

Archaea and bacteria together are the most ancient of all organisms. They branched off from each other very early in evolution when more complex eukaryotes were still billions of years in the future. Modern archaea more closely resemble the cells that populated ancient Earth than bacteria. Many archaea still live in boiling hot conditions that bathe them in acids and sulfur… and they survive without oxygen. All of these conditions mimic the environment of early Earth. Some bacteria can survive harsh conditions too, but bacteria are much more likely to be the prokaryotes you find in your immediate surroundings.

Archaea

Archaea live in so many forbidding environments of acid, searing heat, and noxious gases, they have become synonymous with the word "extremophile." Although archaea dominate life in Earth's extreme environments, they also show up in temperate places where people live.

Archaea look just like bacteria in a microscope. These two domains differ mainly by (1) the types of lipids archaea use to make their membranes and (2) the fact that their membrane consists of a single layer rather than the bilayer common in the rest of biology.

Both features of archaea help them survive in conditions where most other organisms cannot live:

> ➤ Archaea's membrane lipids are composed of very long hydrocarbon chains plus a large lipid-like molecule called squalene. Both of these substances keep the membrane from disintegrating in very hot temperatures.

> ➤ The single layer of archaea would seem to offer less protection to the cell than a bilayer. But this layer contains many lipid strands that tether the membrane together. This configuration also provides protection against a hazardous environment.

With their protective membrane intact, archaea thrive in places you probably thought nothing could survive. For example, archaea populate the hot steam vents at the ocean floor under extremely high pressures. They also enjoy boiling hot sulfur springs, salt lakes, and the inside of volcanoes. Archaea have been found growing in concentrated sulfuric acid! On occasion, archaea join other microbes in the less demanding conditions of soil and water. Let's meet some important groups of archaea.

Sulfur Users

Archaea are crucial to sulfur recycling on Earth. By their actions, sulfur gets converted into chemicals that other organisms can use in their own metabolism. Many sulfur-rich environments are also very acidic or very hot. Archaea recover the sulfur from these places.

The sulfur-using archaea have genus names like *Thermococcus* and *Sulfolobus*. "Thermo-" hints at the microbe's desire to live in hot places. The prefix "sulfo-" refers to sulfur.

Methane Producers

Another word for a methane producer is "methanogen." These microbes constantly remove excess carbon dioxide from the environment and convert it to methane for energy.

Biology Tidbits

The Methanogen Problem

Methanogens take large amounts of carbon dioxide out of the atmosphere. Thank you, methanogens, for helping slow global warming, which is associated with carbon dioxide levels in the air! But after methanogens absorb carbon dioxide, they turn it into methane and release it back into the atmosphere. This gas has several times the global warming capacity of carbon dioxide. This is why scientists think that methane is a bigger threat in climate change than carbon dioxide. More troubling, methane comes from sources that people cannot control. Methanogens emit methane from underwater sediments, landfills, and the digestive tract of cattle, wildlife, and insects.

Methanogens cannot survive in the presence of oxygen. Their metabolism probably resembles the metabolism of ancient prokaryotes that existed in an oxygenless atmosphere with lots of carbon dioxide. Modern methanogens still confine themselves to these anaerobic habitats. They live in very deep soils, the sediments under lakes and rivers, and inside animal and insect digestive tracts.

Example methanogen genera that are also archaea are *Methanococcus*, *Methanobacterium*, and *Methanosarcina*. The term "Methano-" gives you a big hint about the type of metabolism used by these microbes.

Halophiles

Halophiles live only in places where the salt concentration is extremely high. These archaea inhabit salt lakes, brine deposits, and salt mines. Even ocean water is not salty enough for these microbes.

Halophiles have a special system for constantly pumping salt out of their cell so that it does not interfere with normal cell chemistry.

The prefix "Halo-" stands for salt. Typical halophile genus names are *Halobacterium* and *Halococcus*.

Archaea without Cell Walls

A cell wall protects archaea the same as it works in bacteria. The strong cell wall lies outside the membrane and protects the cell from physical harm. Yet some prokaryotes exist without a cell wall. *Thermoplasma* and *Ferroplasma* are two of them. The membranes of these archaea contain high levels of the sugar mannose that perhaps strengthens the membrane.

Thermoplasma and *Ferroplasma* live in active volcanoes and also in the refuse piles near coal mines. *Thermoplasma* prefers the heat generated in coal piles, which can reach 140°F, and acidic conditions. These two specialties make this microbe a thermoacidophile.

Ferroplasma lives in the caustic and acidic runoff from ore mines. The microbe processes the iron compounds that are abundant in these places and also converts sulfur compounds. Unfortunately for our environment, *Ferroplasma* converts sulfur to sulfuric acid, which makes the mine runoff even more dangerous for the soil and streams.

Biologists have put *Ferroplasma* to work extracting certain metals from poor-quality ores.

Biology Vocabulary

Thermoacidophiles

Some microbes are extreme versions of extremophiles! Thermoacidophiles survive conditions of both very hot temperatures ("thermo-") and acidic conditions ("acido-"). They evolved this way because of the advantages their lifestyle gives them. Thermoacidophiles can dominate environments that sustain few other living things.

Bacteria

Bacteria have numerous activities that improve our environment and health. They also include certain pathogens, so perhaps you have a hard time deciding whether to love bacteria or fear them. (Archaea have no known human pathogens.)

Bacteria cover almost every surface and participate in every biological system, and the percentage of pathogens is miniscule compared with the number of bacteria that help us. You can now feel confident in thinking about bacteria as vital and necessary partners with the life on Earth.

Bacteria are the same size as archaea; they average about 1-3 micrometers wide and 2-5 micrometers long. Like archaea, bacteria come in a variety of shapes. Many bacteria live in anaerobic places just like archaea and a few species even withstand very hot temperatures.

Many bacteria have one or more flagella. These tails help the cell swim through water to find food and seek comfortable environments.

Almost all bacteria have cell walls. The bacterial cell wall contains the fiber-like substance peptidoglycan. Bacteria are the only organisms that make and contain this compound.

The bacterial world is thought to be the most diverse group of organisms on Earth. Biologists confront many challenges in figuring out how many species exist. The numbers of bacterial species are only estimates and may be quite inaccurate for the following reasons:

➤ Bacteria live everywhere and live in environments impossible to reach for taking a sample.

➤ Bacteria share genetic material, making the identification of species a difficult task.

➤ Bacteria mutate quickly so that new species arise and some species disappear in short periods of time.

For these reasons, biologists estimate that they have information on less than 1 percent of all bacteria on Earth. Biology has solid information on only about 5,000 species and partial information on another 5,000 proposed species. If these 10,000 species represent as much as 1 percent of all bacteria on Earth, that means the planet holds about one million species of bacteria!

The actual numbers of bacterial cells on Earth are no less daunting. Only recently have biologists come up with estimates of the number of bacteria. The numbers are so big they are hard to comprehend. The number reaches five nonillion. That's five times 10 with 30 zeroes behind it.

Most of these bacteria reside in soil. A teaspoon of soil contains billions of bacteria. The oceans and freshwaters also hold enormous amounts of bacteria. Only the air is relatively sparse in bacterial cells, but the numbers increase if the air also holds dust, soot, and other particles upon which bacteria ride.

Inside a Cell

Peptidoglycan

Peptidoglycan is a very long, continuous molecule called a polymer. Only bacteria make peptidoglycan. They deposit the polymer on the outside of their cell as they build a protective cell wall. The fully formed cell wall contains a distinct peptidoglycan layer. In this layer, peptidoglycan forms an interconnected web of polymers that are strengthened by protein bridges. This peptidoglycan creates one of nature's strongest substances. It allows bacteria to withstand spinning, jostling, crushing, heating, and freezing. Eukaryotic cells that lack the peptidoglycan-based covering succumb to physical forces much faster than those hardy bacteria.

Biologists can group bacteria according to the main jobs the bacteria perform. Keep in mind these are large, diverse groups and that many bacteria have other roles distinct from these main groups.

Photosynthetic Bacteria

Photosynthetic bacteria live in fresh and marine surface waters and any place on land that is exposed to light and offers a little moisture. Therefore, these bacteria cover rocks, plant surfaces, and even grow on houses.

Three main phyla make up the photosynthetic bacteria: Chloroflexi, Chlorobi, and Cyanobacteria. The "Chloro-" prefix refers to the photosynthetic pigment chlorophyll these microbes contain.

Cyanobacteria may be the most numerous photosynthetic bacteria on Earth. Along with algae, cyanobacteria put more oxygen into the atmosphere than all green plants.

Cyanobacteria also have another important role on Earth by absorbing nitrogen gas from the atmosphere. This ability helps keep nitrogen cycling through the atmosphere, soil, and water. Because nitrogen availability can be limited in many environments, plants and animals depend on microbes that can take it from the air and convert it to a form the higher organism can use.

Proteobacteria

Proteobacteria is the largest single group of bacteria and the most diverse. These bacteria help cycle nutrients in ways similar to nitrogen reactions performed by cyanobacteria. Proteobacteria also include some pathogens, bacteria that help digest nutrients, species that break down wastes, and others that recycle sulfur.

Proteobacteria outnumber other bacteria in your digestive tract. In this habitat they help break down food to a form you can absorb through your gut lining. Of course, the fact that proteobacteria live in the digestive tract of humans and other animals means they can also contaminate nature. When animal wastes contaminate food or drinking water, the consequences are usually due to proteobacteria. You might call the diarrhea, headache, vomiting, chills, and fever "food poisoning," but the correct term is "foodborne illness" or "foodborne infection." Proteobacteria cause most of the thousands of foodborne illnesses that occur every year in the United States.

Names that might be familiar to you and which belong to the proteobacteria are: *E. coli*, *Salmonella*, *Shigella*, *Campylobacter*, and *Vibrio*. All of these cause foodborne illness. Other non-foodborne proteobacteria participate in nitrogen recycling (*Rhizobium*) and sulfur recycling (*Beggiatoa*).

Spirochetes

Spirochetes produce spiral-shaped cells. The "Spiro-" prefix hints at the bacteria's shape.

Spirochetes live in the digestive tracts of mammals, mollusks, and insects. Some spirochete genera are important in human health because they cause infections carried either by insects or in sexual *transmission*.

Some spirochete names to know are *Borrelia*, *Leptospira*, and *Treponema*.

Purple and Green Sulfur and Nonsulfur Bacteria

This group consists of four subgroups that differ by the pigments they contain and the ability to use sulfur in their energy reactions. These bacteria also perform photosynthesis but their type of photosynthesis produces no oxygen.

The four main groups in this category are:

> ➤ Purple sulfur bacteria
>
> ➤ Purple nonsulfur bacteria
>
> ➤ Green sulfur bacteria
>
> ➤ Green nonsulfur bacteria

Purple bacteria actually range from pinkish red to brown. Their colors come from a mixture of chlorophyll-like pigments. Green bacteria also contain chlorophyll-like pigments that color them yellowish-green to green.

Purple and green bacteria live in layered mats that grow in certain waters, such as ponds. These mats contain several other types of bacteria that all work together to exchange nutrients.

Mycobacteria

Mycobacteria are important because they cause lung infections in humans and other animals. In humans, the disease is tuberculosis (TB).

Mycobacteria succeed by living inside cells of the infected organism's respiratory tract. A cell wall of unique composition enables this pathogen to avoid the body's immune system and stay in the lungs for a long time. Mycobacteria can persist in the body for decades if an infected person is left untreated.

The species that causes TB in humans is *Mycobacterium tuberculosis*.

Mycoplasma

Mycoplasma has a similar name to mycobacteria, but these two groups are unrelated.

Mycoplasma lacks a cell wall. Like the archaea that lack cell walls, mycoplasma has special lipids in its membrane that help protect the cell contents.

This group contains two main genera: *Mycoplasma* and *Ureaplasma*. These microbes cause infections of the respiratory tract and the urinary-genital tract, respectively.

CHAPTER 19

Algae, Protists, and Fungi

In This Chapter

➤ Characteristics of algae, protozoa, fungi, and yeasts

➤ The main groups of algae and fungi and their importance

➤ Protozoa's role in the environment

➤ The features of a mushroom and mold body

In this chapter some of the eukaryotes step forward to introduce themselves. These are the simple eukaryotes that often live as single cells: algae, protists, and fungi. Fungi can grow into gargantuan multicellular organisms but they also include yeasts. Yeasts are single-celled organisms that behave a lot like bacteria when grown in a test tube.

Algae, protists, and fungi consist of the complex cells that characterize members of Domain Eukarya. These organisms all have membrane-enclosed organelles, a nucleus, and DNA organized into chromosomes.

All three groups contain a diverse collection of organisms that sometimes seem nothing alike. Algae and fungi are especially diverse. They include members that are unicellular (single-celled), multicellular, and filamentous. Protists are usually unicellular but their physiology varies. Protists include aerobes and anaerobes, photosynthetic and nonphotosynthetic organisms, and pathogens and nonpathogens.

Algae

You could soon wear out the word "diversity" when talking about algae. These organisms span a staggering range in physiology, size, and appearance.

The green stuff growing in ponds, a birdbath, or even your fish tank is probably algae. Some of the green sheen on coastal rocks sprayed by the ocean could well be algae. The giant kelp forests that grow along the world's coasts and in aquarium tanks are also algae.

Some of the features of algae tempt biologists to classify them with the protists. These are the algae that live as a single cell and swim around in water environments. Other algae have been proposed as belonging with plants. Certain brown algae and red algae—you know them as seaweeds—seem more like plants than algae.

The normal taxonomy hierarchy into which most living things fit doesn't always work well for the diverse algae. Algologists (people who study only algae) prefer to put algae into seven divisions. The names are tongue twisters but you'll notice from the following list that they go by very familiar common names:

➤ Chlorophyta—Green algae, these organisms are often unicellular and make a significant contribution to global photosynthesis. They live in all waters and on land.

➤ Charophyta—Called stoneworts or brittleworts, only about 250 species have been studied. These algae stay in freshwaters and *brackish* water.

➤ Chrysophyta—Brownish or gold-brown algae, this group includes diatoms, which have unique single-celled structures highlighted by a hard shell of silica. Diatoms are harvested from the ocean and sold as diatomaceous earth.

➤ Euglenophyta—The main member of this group is single-celled *Euglena*, a photosynthetic water organism that swims with a flagellum and serves as a common model for generations of biology students.

➤ Phaeophyta—Called brown algae, this group includes the massive kelp forests in cool-water coasts of California, Alaska, and British Columbia.

➤ Pyrrhophyta—This group goes by the term "dinoflagellates," which means their bodies consist of two segments ("dino-") and they get around in the water using flagella. Some of these algae are responsible for algae blooms and red tides.

➤ Rhodophyta—Red algae, this group includes many seaweeds that collect filtered light for photosynthesis at depths that other photosynthetic organisms cannot tolerate.

Biology Tidbits

Harmful Algae Blooms (HABs)

HABs are dense growths of billions of algae that appear very quickly in water when the water gets a sudden influx of nutrients. Imagine starving on a deserted island and then being rescued by a cruise ship. The captain leads you to the buffet table where you see more food than you have seen in a month. You might well gorge yourself in this situation. HABs start the same way. Small numbers of algae subsist in nutrient-poor waters. Then it rains and a torrent of runoff brings nitrogen, phosphorus, carbon, and other goodies from farms, manure piles, and yards. The algae burst into frantic growth. This burst of growth called a bloom causes problems in the environment.

The algae in HABs produce toxins. The large numbers of cells can produce enough toxin to harm fish and shellfish. If you eat contaminated seafood, you will likely get sick too. Some HAB toxins cause serious neurological disorders, including paralysis. Marine mammals fall victim to HABs the same way as humans. A second hazard of HABs relates to the frenzied cell growth. The algae grow too fast for their own good and die almost as quickly as they grew. Bacteria then start another burst of growth when they start chewing up the delectable algae. The bacterial bloom robs the water of all oxygen and marine life in the area will suffocate if they don't flee. HABs therefore start serious chain reactions in human and environmental health.

The divisions described above include almost every type of plant-like body you can imagine. Cells range from little balls that drift aimlessly in currents to more industrious swimmers that power through their habitat with rotating flagella. Algae also form long filaments, plant-like aquatic bodies that secure themselves to the floor, and beautiful balls made of many single cells joined together. The organism *Volvox* provides an example of one of these multicellular balls that may represent a step in evolution from single cells to more complex organisms.

Protists

Many protists seem like unicellular algae, and vice versa. The alga *Euglena* fits in with protists almost better than it fits with algae.

When I speak of protists, I mean mainly the protozoa. Protozoa are unicellular eukaryotes that live almost exclusively in water and cannot perform photosynthesis. (Good-bye, *Euglena*!) Biologists are fond of saying that "a world exists in a drop of water." This world of life seen in a microscope is predominantly protozoa. Take a shot glass-sized sample from a pond, a marsh, or a intertidal pool. If you can borrow a microscope to study your find, you might be entertained for an hour or more by the cells that dash and dart, twirl and spin, race ahead or back up, or crawl like otherworldly creatures barely seeming to move at all. I could use the word "diverse" again, but I think you get the point.

Protists belong to orders, families, and genera like other organisms. They can also be conveniently placed into categories based on their cell body type. Here are some of the most important types:

➤ Amoebae—Includes the genus *Entamoeba*. These are slow-crawling cells with no definite shape. They live in nature and in the digestive track of animals. Most cause no harm, but *E. histolytica* can be trouble; it causes dysentery.

➤ Sporozoa—These protozoa have life cycles that include different forms of the cell. The sporozoa get their name from an infectious form called a sporozoite. This group includes *Plasmodium*, the cause of malaria. Mosquitoes carry *Plasmodium* and insert it into the bloodstream when they bite.

➤ Ciliates—These cells get their name from the thousands of tiny hairs that cover them. Ciliates beat their cilia in coordinated waves to swim forward, reverse, and change direction. In a microscope, they whiz past faster than you can catch. *Paramecium* is a famous ciliate that lives only in freshwater. *Paramecium* constantly moves and eats by beating its cilia in the direction of a mouth—the mouth is just a crevice called the oral groove. The cilia sweep food particles into the groove. Bacteria often become food for *Paramecium*.

➤ Flagellates—Flagellates contain sub-groups that can be part of other classifications. For instance, dinoflagellates are often claimed by the algae. A group of flagellates called apicomplexans also includes the sporozoan *Plasmodium*.

Biology Vocabulary

Apicomplexans

Apicomplexans lives as a parasite in humans and other animals. As a parasite, it damages the host body that it lives inside. Apicomplexans gets its name from the word "apex," which refers to one end of the sporozoite that penetrates the body's tissue. The term "complexans" refers to the cluster of organelles, or complex, in the apex end of the sporozoite.

For a long time, scientists thought protists played a minor role in the environment. Protozoa caused various debilitating diseases, but the benefits of protozoa remained unknown. Studies on protozoa also lagged behind research on bacteria. Protozoa do not grow in labs as easily as bacteria. Give bacteria enough nutrients and set the incubator temperature to a comfortable level, and bacteria reproduce with abandon. Protozoa proved to be much more finicky. We now know some of the benefits of protozoa in nature:

➤ Protozoa inside the digestive tract of animals and insects help digest dietary fiber.

➤ Protozoa aid in the breakdown of dead plant material in the environment.

➤ Protozoa graze on bacteria in densely populated habitats, such as ponds, thus keeping bacterial numbers under control.

➤ Protozoa serve as food for other tiny organisms, which then serve as food for a chain of larger organisms.

➤ The bodies of dead protozoa sink to the bottom of waters and provide nutrients to the microbes in those nutrient-limited places.

Slime Molds

Slime molds are a unique type of protozoa that were once thought to be fungi. These organisms fooled biologists for a hundred years because these protozoa produce fungus-like bodies and release spores. Slime molds, nonetheless, are not related to bread mold or any other type of fungus.

Slime molds are now known to have fascinating life cycles that make them half-aquatic and half-terrestrial. Within a slime mold life cycle, a species can live as a single-celled organism, and then group together into a multicelluar thing that crawls around on the soil, almost like an animal!

In certain environments, slime mold cells signal each other to swim towards a common location. There, up to 20,000 cells coalesce into a single body that can move. The slime mold heads off across its terrain in search of food. Meanwhile, specialized cells glide around inside the slime mold colony and destroy any bits of garbage or infecting bacteria they might find. If they find a dangerous particle, they grab it and move to the outside of the colony. The specialized cell falls off the slime mold colony and takes the offending particle with it. This sacrifice keeps the rest of the colony safe.

Biologists are beginning to find evidence that slime molds were among the first organisms to physically crawl from the water onto land. Today's experiments show that a slime mold found somewhere in your backyard might be gone if you come back an hour later to check on it.

Biology Vocabulary

Heterotrophs

Organisms that live by heterotrophy require a diet containing a variety of organic compounds. Heterotrophs get the organic compounds they need by eating living organisms or their by-products. Humans are heterotrophs. By contrast, *Euglena* is not a heterotroph because it makes its own organic matter through photosynthesis.

Fungi

Fungi are heterotrophs. This means they must feed on other biological matter to get their nutrition. Fungi cannot make their own food as plants and algae can. For this reason, fungi are not plants.

Most fungi are multicellular organisms for the majority of their life cycle. When fungi exist only in a single-celled form, we call them yeasts.

Multicellular Fungi

Multicellular fungi have a wide variety of body types. Fuzzy molds, frizzy mildew, and mushrooms of every variety belong to this group of organisms.

Multicellular fungi have certain features in common with each other:

➤ Reproductive structure—This varies quite a bit whether we are speaking of mushrooms or molds. The structure is usually aboveground and has distinct coloring and shape that alone can provide enough information to identify the organism's genus.

➤ Hyphae—Tiny filaments that reach from the inside of the reproductive structure to under the soil or along a surface. Fungi grow by building extensions onto the hyphae.

➤ Mycelium—A tangle of hyphae that absorbs nutrients and water to feed the fungus. The fuzzy mass of mold you sometimes see growing on bread is a mycelium.

Some fungi evolved to feed on living tissue, such as animals. For example, the fungus that causes athlete's foot infection (*Trichophyton mentagrophytes*) lives by infiltrating the skin. It prefers the foot to other parts of the body because of the moisture provided by sweat.

Athlete's foot is an example of a parasite; it harms the thing it lives on while getting a benefit for itself. Many other fungi develop other *symbiotic relationships* in nature that are more beneficial. For example, fungi develop commensal relationships, which benefit one partner, as well as mutualistic relationships, which benefit both partners. Lichens offer good examples of mutualism. In a lichen, a fungus lives with a microbe and both do better than if they lived apart.

Biology Vocabulary

Commensalism and Mutualism

Commensalism is a relationship in which two organisms live in close association (symbiosis) wherein one organism benefits but the other is neither helped nor harmed. A mushroom that draws nutrients from a tree's trunk might represent a commensal relationship if the tree is unaffected.

Mutualism is a type of symbiosis in which both organisms benefit. For example, some fungi attach to plant roots and grow an extensive system of filaments called mycorrhizae. The mycorrhizae supply to the plant minerals that the plant cannot absorb from the soil by itself. The fungus meanwhile gets a benefit when the plant supplies it with organic compounds.

Multicellular fungi also have complex life cycles that often contain an asexual phase and a sexual phase. But there are lots of exceptions in all of biology! Some fungi have only asexual reproduction and others have only sexual reproduction.

Biology Tidbits

Lichens

Lichens are made of two organisms living in symbiosis. This type of symbiosis is mutualistic because both organisms are helped. One partner in lichens is always a fungus. The other partner can be either algae or cyanobacteria. Each organism provides something that the other could not obtain on its own. The fungus provides a safe habitat among the mycelium that emits carbon dioxide. The algae or cyanobacteria provide sugars that they produce in photosynthesis.

More than 13,500 types of lichens exist. Their relationships are so strong, biologists have identified most of these partnerships as lichen species. Lichens, in fact, account for about a fifth of all known fungi.

Mushrooms

You certainly know some of the practical uses of mushrooms: soup, pizza, appetizers, etc. In nature, mushrooms provide an equally important service. Mushrooms excrete enzymes into the environment that degrade dead plant matter. The mushroom enzymes are especially effective in breaking down the strong plant fibers lignin and cellulose. Few other things in nature can degrade lignin.

Mycologists (people who study mushrooms) divide mushrooms into two main groups:

➤ Ascomycetes—Nicknamed "sac fungi" because they hold their reproductive spores in a sac called an ascus. Bread mold (*Neurospora crassa*) is an example of an ascomycete.

➤ Basidiomycetes—This group contains mushrooms, rusts, smuts, and some molds. During their life cycle, these fungi produce a club-shaped appendage called a basidium. The basidium holds the reproductive cells. Basidiomycetes are important decomposers of dead trees and wood. Puff balls and the deadly mushroom *Amanita muscaria* (nicknamed the "death cap") offer examples.

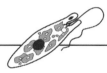

Inside a Cell

Filamentous Growth

Fungi grow by extending hyphae and expanding their mycelium. Filamentous growth helps fungi survive in their environment, especially since fungi are otherwise immobile. The spreading of hyphae enables the fungus to reach new nutrient sources. Filaments grow in between soil particles, over surfaces, and into living and dead tissue. Filaments also allow fungi to form symbiotic relationships with plants and microbes.

Molds

Molds are powerful decomposers of organic matter. You probably never think about molds until you discover their growth covering a leather belt or something in the refrigerator that used to be food.

Molds use both asexual and sexual parts in their life cycle. At certain steps in the life cycle, the mold produces a spore. Depending on the life cycle stage, a spore is either haploid or diploid.

The wind carries spores for distances of up to a mile. Many molds help the spores on their journey by building structures that launch the spores airborne with considerable force. The spores eventually land and then germinate in soil. The spores that have not yet landed might find their way up your nose and cause an *allergic reaction*.

Yeasts

Yeasts live in liquids and moist places. They are asexual and reproduce by budding in which a small daughter cell grows out of the parent cell. The bud drops off from the parent when it has reached close to full size. This leaves a bud scar on the parent cell.

I said that yeasts are unicellular, and that is true most of the time. In a few species, however, yeasts can form filaments and in a microscope they look like other filamentous fungi. The availability of specific nutrients determines whether a yeast will temporarily turn into a filamentous form.

Humans have made good use of yeasts since the earliest recorded history. Beer brewing, winemaking, and bread making are ancient enterprises that all use yeasts. *Saccharomyces* is the workhorse of the yeast world. The species *S. cerevisiae* helps in the fermentation process used in alcohol production and in bread making. In breads, the yeast is valued more for the carbon dioxide it emits rather than alcohol production. The carbon dioxide produces bread's characteristic bubbles and makes the bread rise.

CHAPTER 20

 **Viruses**

In This Chapter

➤ The features of viruses and parts of a virus

➤ How scientists classify viruses

➤ The basics of a virus infection cycle

➤ An introduction to viral diseases

In this chapter you will learn about viruses. Biology is the study of living things, but viruses do not fit neatly into that definition. Viruses cannot exist on their own; they must be inside a living cell to survive. For this reason, some biologists (myself included) do not refer to viruses as "organisms." Instead, try to think of viruses as particles.

Why do viruses exist in a gray area between the living world and nonliving things? Viruses resemble living things by having genetic material, either DNA or RNA. Viruses also replicate. Replication is a major attribute possessed by living things and lacking in nonliving things. But viruses cannot live on their own or replicate on their own. They don't take in nutrients and they cannot sense their environment or respond to it. This makes them more like a speck of dust than a microbe. Yet viruses are usually included with the microbes and, whether living or not, viruses have a tremendous effect on the biological world.

Viruses are obligate parasites. They cannot exist in any other way but to get inside a living cell and commandeer the cell's machinery. Viruses replicate by making the cell do all the work! Like a political coup, a virus slips across the border (the plasma membrane) and enters the capital (the nucleus). It then infiltrates the highest seats of government (the cell's DNA). The takeover is complete when the cell's replication system makes more viruses instead of managing cell reproduction.

Viruses infect archaea, bacteria, plant cells, and animal cells. Surely, such an agent must have had an important role in evolution. Viruses continue to impact all living things today. They do this with one of the simplest structures in all of biology.

Meet a Virus

A virus particle consists of three parts. The first is the viral genetic material. Some viruses carry DNA and others carry RNA, but no virus carries both. The second piece is the capsid. A capsid is a shell made of tightly woven protein that encloses the genetic material. Finally, some viruses have an outer coating high in lipids. You will soon see that this so-called lipid envelope changes a virus' ability to infect.

The parts of a virus collectively make up the virion, which is an entire virus particle.

Viruses come in a variety of shapes that are characteristic of the particular virus. For example, influenza virus always looks like a slightly misshapen ball with lollipop appendages sticking out all over it. The rabies virus is always in a cylindrical bullet shape.

Viruses that infect mammals and other animals come in about a dozen capsid shapes or configurations of the genetic material inside the capsid. The shapes are so distinctive, they can be used as a way to classify viruses.

The only truly unique shape occurs in bacteriophages, which are viruses that infect bacterial cells.

Hundreds of thousands of viruses are thought to infect every living thing. Some viruses attack only one species but other viruses, such as influenza, infect more than one. The viruses that infect plants do not usually infect animals, but some exceptions exist even to this rule. Some species are known to be infected by at least 100 different viruses. Fortunately, they don't all infect at the same time!

Bacteriophages: Viruses that Attack Bacteria

Biologists sometimes refer to bacteriophages as simply "phages." Phages have a unique structure compared with other viruses. The capsid has a polyhedral shape (many-sided) like several other virus classes, but the capsid attaches to a second cylindrical piece called the tail. The tail holds several inverted "V" legs that make the entire phage look like a lunar landing craft.

Because phages are deadly to bacteria but leave humans alone, virologists (people who study viruses) use phages in lab experiments in place of more dangerous viruses.

Biology Tidbits

Tobacco Mosaic Virus (TMV)

TMV is one of the most studied viruses in virus research. Perhaps that is fitting since TMV was the first virus to be discovered. In 1883, German scientist Adolf Mayer discovered that disease in tobacco plants was transmitted by the extract from diseased plant leaves. Nothing showed up in Mayer's microscope, so he concluded he had found a rare extra-small type of bacteria.

About 10 years later, Russian scientist Dimitri Ivanowsky delved further into the disease-causing extract from tobacco. He passed the liquid through a filter of tiny pore sizes that would remove all bacteria. When Ivanowsky injected the filtered material into plants, they still got diseased. Next, Dutch botanist Martinus Beijerinck injected the filtered extract in a series of tobacco plants. After they became diseased, he collected extract from these plants. This extract could also cause disease. Beijerinck concluded that the mysterious invisible particle had reproduced inside the plants. Though too small to see in a microscope, something biological seemed to exist in the extract. Not until 1935 did technology develop to enable American Wendell Stanley to discover the particles in the plant extract: TMV.

Virus Size

Bacteria are microscopic; they are measured in micrometers. Viruses are submicroscopic and virologists measure them in nanometers. A nanometer (nm) is one-billionth of a meter.

Viruses range from about 10 nm to 300 nm in diameter. Cylindrical viruses can reach to almost 1,000 nm in length but only 10 nm in width. Polio is one of the smallest viruses that infects humans at 30 nm in diameter. The Ebola virus is a chunky cylinder of about 970 nm in length.

Regular light microscopes that work fine for studying bacteria cannot show us a virus particle. Virologists use electron microscopes to study virus shapes. These instruments cost close to a million dollars and require special training to use.

How We Classify Viruses

Virus classification does not follow the same type of taxonomy used for living organisms. Different classification schemes work well in virology:

> ➤ Type of genetic material: DNA or RNA.

> ➤ Method of replicating inside the cell.

➤ DNA or RNA size.

➤ Presence or absence of the lipid envelope.

➤ Capsid shape and symmetry.

➤ Main organism it infects.

➤ Type of disease it causes.

➤ Method of transmission.

The dozen or so types of viruses that infect humans have been assigned to their classes based on capsid, genetic material, and replication method.

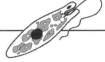

Inside a Cell

Influenza and Antigenic Drift

Why does influenza (the flu) come back into the human population every year and cause a large number of illnesses? Despite the widespread use of the flu vaccine, influenza always seems to get around our defenses.

Influenza is one of many viruses that regularly rearranges its genes. Often, two or three different strains of influenza mix their genetic material. This action invents a brand new influenza virus never before seen. The periodic reassortment of genetic material is called antigenic drift. Antigenic drift causes your body to be forced to recognize a new version of the virus each flu season. This makes your immune system much less effective at fighting infection than if the same influenza showed up every year.

How Viruses Infect

Virus infection follows a cycle of five general steps:

➤ The virus enters the cell and quickly sheds its protein coat to expose its genetic material.

➤ The cell's enzymes start to replicate the virus' genetic material.

➤ The cell also incorporates viral genes into its own DNA. Every time the cell replicates its DNA, it actually produces more viral capsid proteins and genetic material.

> ➤ The new virus pieces self-assemble inside the cell cytoplasm.

> ➤ New virus particles leave the cell and seek other cells to infect.

This general scheme can produce hundreds to thousands of new virus particles from one infected cell. The number of particles made depends on the particular virus that has infected a specific organism.

Viruses use two versions of the general infection scheme. One is the lytic cycle and the other is the lysogenic cycle or lysogeny.

The Lytic Cycle of Virus Infection

The lytic cycle follows the general infection scheme described above, but at the last step the virus destroys the cell. When hundreds of virus particles are ready to burst free from the cell they so rudely used, they do just that… they burst. The cell lyses, meaning it breaks apart. The lytic cycle therefore leads to cell death.

The Lysogenic Cycle of Virus Infection

Viruses take a more insidious approach to infection when they use lysogeny. In this cycle, the viruses do not lyse the cell upon exiting. Instead, the virus lets the cell package it up in an envelope made of a portion of the cell membrane. The packaged virus gently pops off the cell like a bubble and the cell lives another day.

Lysogeny can continue for many cycles. All the while the cell manufactures and ships new batches of virus with an intent to spread infection.

Inside a Cell

What Makes HIV Dangerous

Cells have defenses to protect against the very type of infections that viruses cause. Enzymes inside cells chop foreign DNA into pieces. Some cells eject foreign invaders before they can cause harm. But viruses have been around since the earliest cells, so they also have a few tricks to help their survival.

The most difficult type of virus to defend against may be the virus class called retroviruses. Retroviruses carry RNA and an enzyme that converts RNA into DNA. This action is in reverse to what normally happens in living things. Normally, DNA goes to RNA, which goes to proteins. This is why we call these viruses "retro." By working the system in reverse, retroviruses confuse the cell's defenses.

The human immunodeficiency virus (HIV) that causes acquired immunodeficiency syndrome (AIDS) is a retrovirus. HIV employs other features that add to its danger. First, it hides inside the body's immune system, which is the very system meant to kill infectious agents. Second, HIV changes its outer surface to keep the body's immune system from recognizing it. Finally, HIV is a latent virus. This means it can hide silently in the body for decades before causing disease.

At some point in lysogeny, the virus decides it has had enough and switches over to the lytic cycle.

Viral Diseases

A virus does not infect a cell without causing significant problems for the cell. In the lytic cycle, viruses destroy the cells they infect. This is part of the viral disease.

Because viruses are closely associated with the type of tissue they invade, viruses can be classified according to the diseases they cause:

> ➤ Upper respiratory tract—cold (rhinovirus); flu (influenza).

> ➤ Lower respiratory tract—Viral pneumonia (respiratory syncytial virus).

> ➤ Nervous system—Rabies (rabies virus); AIDS (HIV); viral encephalitis (encephalitis virus).

> ➤ Urinary-genital tract—Genital warts (human papillomavirus).

Viruses sometimes invade more than one tissue type over a long course of infection. For example, the herpes virus causes cold sores, blisters, and a rash on the skin. Herpes is also classified as a sexually transmitted disease, so its main infection route is via the genital-urinary tract.

Other Nonliving Particles in Biology

Two particles that are not viruses invariably get grouped with viruses because biologists have nowhere else to put them! They are viroids and prions. These particles have very simple structures and also do not live independently.

Viroids

A viroid is a naked piece of RNA that infects plant cells. No capsid protein surrounds and protects it. The RNA folds in a specific manner that protects it and also exposes an infective end.

Viruses and viroids cause significant losses to plant crops each year. The main plants attacked by viroids are potatoes, avocados, coconuts, chrysanthemums, and some citrus species.

Prions

A prion is a naked piece of protein that infects animal cells. Prions were discovered in 1982 and stunned biologists who could not believe a protein strand could behave like a virus.

Proteins, after all, are nutrients we need. How could a protein cause infection? Prions were furthermore tied to very serious neurological diseases in humans and animals.

Prions cause a set of related diseases associated with a gradual decomposition of brain tissue. Humans can get prions by eating infected meat. Thankfully, prion infection is much rarer in humans than in animals. These diseases have different names depending on the infected species, but they all behave in a similar way in the body. The prion neurological diseases are:

➤ Humans—Creutzfeldt-Jakob disease, kuru, Gerstmann-Sträussler-Scheinker syndrome, and fatal familial insomnia.

➤ Cattle—Bovine spongiform encephalopathy, also called mad cow disease.

➤ Sheep—Scrapie.

➤ Deer, elk, and moose—Chronic wasting disease.

➤ Lions, tigers, cheetahs, ocelots, mountain lions, and domestic cats—Feline spongiform encephalopathy.

PART SIX

Complex Animals

CHAPTER 21

 **Invertebrates**

In This Chapter

➤ A review of the characteristics of early animal life

➤ Introduction to the main groups of invertebrates

➤ Non-animal-like and animal-like invertebrates

➤ A special look at sponges

In this chapter we will move along the tree of life from prokaryotes and simple eukaryotes to more complex organisms. All of these are eukaryotes.

Invertebrates are animals without a backbone. They are much simpler in structure and size than the vertebrates, which include humans, but they also have many things in common with what we call "higher organisms."

Invertebrates are multicellular and their nutrition is heterotrophic. This means their food must supply a variety of organic compounds. Like higher organisms, invertebrates develop from an embryo, and the embryo's tissue differentiates into the many types of tissue that form the adult.

The animal kingdom has 35 *phyla*. Twenty-four of these phyla are invertebrates. These animals are numerous, diverse, and outnumber all other animals.

This chapter covers the characteristics that invertebrates developed to separate themselves from prokaryotes and the simplest eukaryotes. I will then introduce you to some of the important invertebrates on Earth.

Characteristics of Early Organisms

Prokaryotes and algae, protists, and fungi follow a free-form type of development when they grow. It is true that bacteria of the same species always have the same shape and size, but their insides are a mess! You would be equally hard-pressed to find an alga, protist, or fungus that adheres to a rigid body plan. Protists such as amoeba have no set shape at all. Is a fungus supposed to have 100 hyphae, 1,000 of them, or a million? No one knows.

Animal life starting with the invertebrates is built on a standard plan. This plan always has a few basic requirements, like building a house. Houses come in many shapes and sizes, but the basic plan should always include rooms, walls, a roof, kitchen, and a bathroom. In animals, the plan has five main requirements:

> ➤ Symmetry.

> ➤ Tissues of different types.

> ➤ Body cavities.

> ➤ Cleavage (No! Not *that* kind!). The pattern of cell division during embryo development in which the zygote splits into two cells, two cells split into four, four to eight, and so on. The patterns and directions of cleavage give rise to symmetrical body parts.

> ➤ Coelom, an inner tube lined from beginning to end with the same type of tissue. The animal digestive tract represents the coelom.

Some animal life bypasses at least one of the above requirements. (Every rule in biology seems to come with exceptions as science discovers more about biodiversity.) The above requirements are furthermore still being discussed and debated among scientists. These debates lead to occasional reclassifying of some organisms from one group into another. Invertebrates nonetheless teach us the basics of how animals developed on Earth.

Sponges: Exceptions to the Animal Body Plan

Sponges are invertebrates that do not follow the rules of symmetry. These are sedentary or sessile organisms, meaning they attach to an immovable surface and stay there. This lifestyle forces sponges to draw food toward them—luckily, they live underwater—by using appendages. Animals that use any derivation of this type of feeding are sometimes called "filter feeders." The more accurate term for sponges is "suspension feeders," which capture food suspended in water as the water passes through their body.

Sponges are hermaphrodites. These are organisms that produce both eggs and sperm cells and conduct both female and male functions in reproduction. That's handy for an animal that doesn't get around much!

A sponge body lacks distinct tissues, but it does contain several different cell types. Cells called amoebocytes meander through the sponge tissue. As you might guess from their name, amoebocytes have no set shape. They squeeze through every part of the sponge in search of food particles. After digesting the food, they carry nutrients throughout the sponge body. Amoebocytes also produce fibers that strengthen the sponge.

Sponges have several commercial values for people. The first obvious value is as a source of the material used for scrubbing things. Sponge species also produce antibiotics and other substances valued in medicine. Recently, sponge secretions have been studied for their anti-cancer properties.

Biology Vocabulary

Symmetry

Symmetry is a characteristic of most higher organisms. These organisms develop on a symmetrical plan in which each half of the body mirrors the other half. Mammals and most other animals have bilateral symmetry like a spoon. Animals with bilateral symmetry always have a top (the dorsal side), an underside or bottom (the ventral side), a head (the anterior), and a tail end (the posterior).

Some animals have radial symmetry, like a bowl. Sea anemones, for example, take this more round shape. They have a top and bottom but lack defined left and right sides.

Invertebrates That Do Not Look Like Animals

Some invertebrates resemble protozoa more than they resemble complex animals. But they are invertebrates because they all possess symmetry, a body cavity, and differentiated cell types.

This group contains six phyla of small and simple organisms:

> ➤ Placazoa—This phylum contains only one species called *Trichoplax adhaerans. T. adhaerans* looks like a large protozoan, but it is made of cells that form a flexible double-layered plate. The entire organism contains only four different cell types and the smallest quantity of DNA ever found in an animal. *T. adhaerans* is hard to find in nature; it measures only about two millimeters (less than a tenth of an inch) end to end. Scientists usually find it in moist habitats in the tropics.

> ➤ Rotifera—About 1,800 species belong to this group distinguished by a simple organism with a large mouth and prominent digestive canal. These organisms spend their days sweeping microscopic bits of food (prokaryotes mainly) into their mouth using cilia that rim the opening.

> ➤ Loricifera—These deep-sea organisms measure less than one millimeter and eat mostly bacteria. A mouth is rimmed with cilia for taking in food. The phylum contains about 10 species, but many more may exist that have not been discovered.

> ➤ Kinorhyncha—About 150 species make up this phylum. These organisms look like small bugs or big protozoa. Their cylindrical bodies are covered with protective plates and have a mouth at one end. The entire organism measures one millimeter (four-hundredths of an inch).

> ➤ Porifera—These are the sponges of which about 5,500 species exist. Sponges are completely sedentary and wait for food to come to them. Because of their lifestyle, early biologists thought sponges were plants rather than animals.

> ➤ Cnidaria—This phylum includes hydras, jellies, and corals. These organisms have a body cavity accessed by a single opening that acts as both a mouth and anus.

The above phyla offer wonderful study models for seeing how primitive life developed. The serve as food for larger organisms and make up the animal portion of plankton called zooplankton.

Invertebrates That Look Like Animals

Let's go on to the invertebrates that actually make you think of an animal when you see them. These organisms are larger and more complex than the members of the six phyla described above. But the ones listed below are still quite simple compared with vertebrates.

> ➤ Platyhelminthes—This is a phylum of about 20,000 species of flatworms with a central nervous system and sensory structures.

> ➤ Ectoprocta—This phylum has 4,500 species of sedentary organisms called bryozoans. They look a bit like a cross between a sponge and an anemone. They live in aquatic colonies and each body is covered with a tough protective layer called an exoskeleton.

> ➤ Phoronida—This is a small group (about 20 species) of marine worms.

> ➤ Brachiopoda—This phylum has 335 species of marine cold water shells that resemble clams. They differ from clams by having a strong stalk that anchors them to a surface.

> ➤ Nemertea—These are the probiscus worms. You might see one of the 900 species burrowing through sand or swimming through water.

➤ Acanthocephala—More worms! These 1,100 species are thick, segmented, and possess a curved hook at one end. Many infect other invertebrates, such as crabs.

➤ Ctenophora—This phylum (100 species) contains organisms that resemble jellyfish. Called comb jellies, these translucent organisms have eight combs, which are lines of cilia that enable the ctenophore to swim.

Biology Vocabulary

Exoskeleton

An exoskeleton is a deposit of hard material on the outside of an animal's body. A clam's shell offers a good example of exoskeleton. As animals developed and became more complex, their complexity also called for better support for internal organs. The exoskeleton served this purpose. The exoskeleton of arthropods is segmented and attached to underlying muscles. This configuration makes movement easier for the organism. Arthropod exoskeletons are also made of chitin. This very strong material is nearly indestructible in nature; very few enzymes can break it down.

The internal skeleton of vertebrates like mammals is called an endoskeleton.

The following phyla contain members that suggest the word "organism" when you see them.

➤ Mollusca—Snails, clams, squids, and octopuses (93,000 species).

➤ Annelida—Segmented worms, such as earthworms (16,500 species).

➤ Priapula—More worms! With enlarged segment at the anterior end (16 species).

➤ Nematoda—Roundworms found in almost every soil and water habitat (25,000 species).

➤ Arthropoda—Insects, arachnids (spiders), and crustaceans (probably more than one million species).

➤ Cycliophora—Only one species! Discovered in 1995. *Symbion pandora* is a tiny organism that lives attached to the outside of lobsters.

➤ Tardigrada—Small mite-like creatures that live in marine and freshwaters and on moist plant surfaces (800 species).

➤ Onychophora—Worms called velvet worms that live mainly in humid forests (110 species).

➤ Hemichordata—Acorn worms that live in marine mud (85 species).

➤ Echinodermata—Sand dollars, sea urchins, and sea stars (7,000 species).

Chordata

One phylum included with invertebrates contains members that actually have a backbone. This is phylum Chordata, which needs special mention here.

Chordata contains the invertebrate animals called tunicates, lancelets, and hagfishes. Each of these types has four features in common that classify them as invertebrates:

➤ Nerve cord—A hollow column running almost the length of the organism and representing a central nervous system.

➤ Notochord—A long flexible rod that runs between the digestive canal and the nerve cord. This rod is made of fibrous tissue and supports the body much like a backbone does in vertebrates.

➤ Pharyngeal cleft—A slit connected to the mouth that allows water to enter and leave without going through the entire digestive canal.

➤ Tail—A strong extension off the anterior of the body that propels the organism in water.

These adaptations of Chordata invertebrates may seem primitive at first glance, but they represent important steps in evolution. They contribute to refinements of the nervous system, digestion, and mobility.

➤ Tunicates—This group includes sea squirts. These are sedentary aquatic animals that filter water through a netlike structure. Sea squirts get their name from the ability to shoot water at an attacker.

➤ Lancelets—A lancelet is a straight, fairly rigid worm-like creature that burrows almost its entire body into submerged sand. It leaves its mouth exposed at the surface to draw in food using small tentacles. Lancelets do not swim well, but they frequently move their location in search of better feeding. They usually swim toward the water's surface and then slowly sink toward the bottom. When they hit bottom, they continue burrowing into the sand.

➤ Hagfishes—Hagfishes develop a skull made of cartilage and have an anterior end that looks a bit like a head. They have a brain (although it's a small one), eyes, and ears, and a nasal opening that connects with a pharynx. Thirty species of hagfishes exist. All of them live at the bottom of marine waters.

The Chordata species that have a backbone added structures that made them look more and more like animals that think, react, and live in ways similar to how mammals live. Getting to know the invertebrates, however, can help you see the progression of life forms from single-celled organisms to primitive multicellular creatures to sophisticated life.

CHAPTER 22

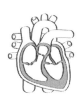# The Animal Body: The Nervous System and Circulation

In This Chapter

➤ Tissue structure and function and humans' tissue types

➤ An overview of body metabolism and your physiology

➤ An introduction to the nervous system

➤ An introduction to the circulatory system

In this chapter you become better acquainted with how the animal body works. Some differences exist between the physiologies of all higher animals, but the body's chemistry in cells remains the same. In fact, all animal bodies have the same systems that make them go in pretty much the same way.

A body system is a collection of organs that work together to carry out a particular function. The digestive system gets nutrients out of food; the immune system fights infections; and the nervous system runs your body's communications.

The systems connect with each other, too. Nerves send messages to muscle; hormones from the endocrine system control the reproductive system; and blood (the circulatory system) carries nutrients to all other tissues. This chapter focuses on two systems that keep all the others working: the nervous system and the circulatory system.

To understand all the body's systems, it helps to know the basic tissue types that make up your body. This chapter begins with explaining tissue types and their structure and functions.

Tissue Structure and Function

Tissue is a group of cells with a common structure and function. The structure of the cells relates to their shape and how the cell is put together. A cell's structure also relates to its function in the body. Biologists call this concept "form follows function." This means that the structure of a cell or body part relates to its job in the body.

Know the form of a cell and you might be able to guess its function. Conversely, if you already know the cell's function, its form suddenly makes a lot of sense.

The body is composed of four tissue types: epithelial, muscle, connective, and nerve.

Epithelial Tissue

Epithelial tissue makes up the skin and the linings of lumens. A lumen is an open space such as the mouth or the intestines.

The cells of epithelial tissue pack closely up against each other. This arrangement helps the tissue carry out its main job: prevention of foreign matter from entering the body.

A specialized type of epithelial tissue called mucous membrane lines the mouth, digestive tract, and genital-urinary tract, and is also in the ear canal and the eyes. This tissue secretes the substance called mucus. (The word "mucus" is a noun for the thick secretion. The word "mucous" is an adjective to describe this type of tissue.) Mucus lubricates surfaces and plays a part in keeping infectious microbes out of the body.

The general term for any layer of epithelial tissue is "epithelium."

In addition to mucous membranes, epithelial tissue comes in three main types:

➤ Glandular—Mucous membranes provide the best example. The cells of this tissue act as miniature glands and secrete something the body uses.

➤ Simple—This epithelium consists of a single layer of cells.

➤ Stratified—This epithelium contains more than one organized layer.

Cell shapes also vary in epithelium in different parts of the body. Cells are cube-shaped (cuboidal epithelium), rectangular (columnar pithelium), or flat (squamous epithelium).

Biology Tidbits

Form Follows Function

Many nerve cells stretch out for more than a foot in the human body. This length is one of the features needed by a cell whose main job is to communicate with other cells. Nerves send and receive messages. A long cell carrying messages like lightning serves you much better than a plump fatty cell that prefers to relax. By contrast, blood cells are compact. Their job includes delivering oxygen throughout the body or going to the site of an infection. This demands a small size so the cell can navigate tiny vessels.

All living things evolved so that form and function make sense in relation to each other. Biologist Charles Darwin showed that working toward the perfect form is an ongoing process. If conditions in the environment change, a population will adapt to that change. Often, the adaptation is a change in form. Populations of finches on various Galápagos Islands proved that form follows function. Darwin noticed that finches on different islands differed in shape and size of their beaks even though they appeared to be the same species. The food supply, however, differed on each island. The finch populations had adapted to the type of seeds that made up the majority of their diet by developing the appropriate tool: their beak.

Muscle Tissue

Muscle tissue is mostly composed of long cells called muscle fibers. Muscle fibers receive signals from the nervous system but they do not send any messages in return.

The main function of muscle tissue is to contract. To do this, muscle contains large numbers of fibers arranged in parallel in bunches. When these bunches get a "Contract!" signal from a nerve, the fibers slide along each other to form what you recognize as a muscle contraction. This takes energy, and muscle tissue demands more energy than almost any other tissue of the body.

The muscle fibers contain high amounts of two specialized proteins: actin and myosin. The connections between these two proteins are responsible for muscle contraction and its eventual relaxation.

You have three forms of muscle tissue: skeletal, cardiac, and smooth. Only skeletal muscle can be controlled by you and your nervous system. This is the muscle of your limbs, torso, and face and head. Contracting these muscles amounts to a voluntary action. By contrast, cardiac and smooth muscle are called involuntary muscle because their contraction is hardwired in your body and you do not have to consciously think about each contraction.

Cardiac muscle runs the heart, and smooth muscle produces a gentle wave along your intestines so that digested food keeps moving.

Connective Tissue

Connective tissue supports other tissues. It tends to consist of protein fibers that form a loose web and are suspended in a jelly-like or soft solid substance.

The body contains three types of connective tissue fibers:

➤ Collagenous—Made of the protein collagen, these are the most abundant type of connective tissue, strong, and nonelastic. One of collaganeous tissue's most important jobs is to keep other tissue connected to bone.

➤ Rectangular—Also made of collagen, these fibers join connective tissue with other tissues.

➤ Elastic—Made of the protein elastin, these fibers offer stretchiness, like rubber, in your moving parts.

The body has several different forms of connective tissue for a variety of purposes. Some connect tissues to other tissues as mentioned above. Also included with connective tissue are bone, cartilage, adipose (fat) tissue, and blood.

Nerve Tissue

Nerve tissue has specialized elongated nerve cells called neurons. This tissue has the ability to build an electrical difference across its membrane. This electrical difference is what creates the signal carried along the nerve fiber to other parts of the body. The inside-outside difference in charge is called an action potential. The nervous system runs on a constant stream of action potentials racing all over your body like an electricity grid.

Inside a Cell

Action Potentials and Resting Potentials

A neuron creates an action potential by pumping sodium, potassium, and chlorine back and forth across its membrane. (For this reason, biologists also refer to the action potential as a membrane potential, mentioned in Chapter 5.) Movement of tiny amounts of positively charged sodium and potassium and negatively charged chlorine results in a pulse that moves rapidly along the nerve. The charged form of these elements are called ions. The ions that move nerve impulses throughout your body are sodium (Na^+), potassium (K^+), and chloride (Cl^-).

When a neuron is not transmitting a signal at any particular moment, the situation is called a resting potential. The resting potential serves as a period in which sodium, potassium, and chlorine can redistribute across the membrane to their original levels. Once they have moved back in place, they are ready to participate in another action potential.

Body Metabolism: How It All Works Together

The body systems' goal is to keep you in a state of homeostasis. Chemists call this a steady state. Think of homeostasis as being in your "comfort zone."

A big part of homeostasis relates to temperature. Various animals have different ways of regulating their body temperature even as the temperature of the surroundings rises or falls. Temperature regulation is a good example of a feedback system: the body senses the outside temperature and adjusts certain processes inside the body to cool it or warm it up. The process of keeping your body's temperature in its normal range is called thermoregulation.

All the systems working together comprise your physiology. Everyone's physiology differs a little. You have probably noticed that in the same room, some people tend to get chilly and other people complain that it's too warm. Perhaps you tend to gain weight easier than your best friend; maybe you cannot keep an ounce on your frame no matter how many cupcakes you eat!

Thermoregulation turns out to be an easy indicator of how different our physiologies are even though we all use the same body plan. Humans and other mammals use an endotherm approach to regulating body temperature. Compare this to a second body type in the animal world, called ectotherms.

Biology Vocabulary

Homeostasis

Homeostasis is a steady state condition of your body. In homeostasis, your temperature is within its normal range and your metabolic rate is appropriate for whatever activity you're doing at the moment.

Ectotherms and Endotherms

Ectotherms are animals that draw heat for their body from their surroundings. These are the so-called cold-blooded animals. When it is cool outside, an ectotherm like a lizard has a slow metabolism and might even be unable to move fast enough to escape a predator. An ectotherm living in its natural climate can cool itself by seeking shade and warm itself by basking in the sun.

Examples of terrestrial ectotherms are lizards, snakes, turtles, and amphibians. Ectotherms of water environments include fish and most invertebrates. They must regulate body temperature by finding warmer or cooler waters.

Endotherms use their internal metabolic rate to heat the body. These are known as warm-blooded animals.

Thermoregulation

The animal body constantly monitors body temperature. Animals can change their body temperature by doing certain activities such as eating, sleeping, and exercising, or through emotional changes. Other factors that contribute to your temperature are age, menstrual cycle, and an infection that causes fever.

When your temperature starts creeping a bit out of its normal range, the body has ways to automatically keep it under control:

➤ Insulation—Skin, fur, hair and other tissues that make up the *integumentary system* help keep body temperature in a constant range.

➤ Evaporation—Most mammals and birds use this technique for cooling. Evaporation by losing water across the skin helps heat escape the body.

➤ Circulatory system adaptations—Many animals control the amount of blood flowing from the body core to the extremities. The action of vasodilation opens vessels wider and allows more heat to move outward. Vasoconstriction makes the vessels narrower and holds heat in the center of the body.

➤ Adjusting metabolic rate—In some mammals, hormones prompt the mitochondria to increase their rate of reactions. The mitochondria produce heat instead of ATP. Shivering is a similar response to cooling by running muscles without doing work. The shivering raises the body temperature.

Biology Tidbits

Hibernation

Some animals evolved with the ability to hibernate over extended periods of cold temperature. In hibernation, the body physiology changes dramatically. Metabolism decreases. This means that the chemical reactions of the body run at much slower than normal rates. The heart rate and respiratory rate also slow down. A hibernating animal relies on stored fat to maintain the slow metabolism. This situation causes the body temperature to decline to lower than normal.

In addition to these automatic ways to thermoregulate, behavior also plays a part in maintaining body temperature. In some animals, hibernation is a way to escape the cold. Some animals migrate to warmer or cooler climates. For people, the act of putting on a sweater is a behavioral response to a change in temperature. The next time your spouse complains because of your fiddling with the thermostat, say "I'm just trying to maintain my homeostasis."

The Nervous System

The nervous system helps you sense and interpret not only your environment's temperature but a variety of other physical features of your surroundings. (It works in close connection with the sensory system, which I discuss in the next chapter.) This system communicates the information about your surroundings to a central control center, the brain, which figures out a suitable response. Your brain does this so naturally you hardly have to wonder why you crave a cup of cocoa on a snowy day or a dive into a pool on a searing summer day.

The layout of the nervous system differs in animals from a central information highway along a spine to the brain, as in humans, to net-like configurations:

➤ Nerve nets—Interconnected nerve cells in a net-like pattern, found in hydras.

➤ Star—A centralized ring holds nerves that radiate from it, found in sea stars.

➤ Nerve cords—Two main cords run the length of the body, connected by tranverse nerves that run laterally, as found in flatworms and the chiton, a type of mollusk.

➤ Squid style—A nerve cell bundle representing a brain holds several cells that extend from it to the rest of the body.

➤ Insect style—A brain-like bundle similar to the squid's connects to a loose arrangement of nerve cells extending throughout the body.

➤ Chordate—A brain connects with a spinal nerve from which other nerves extend.

Neurons and Glia

The neuron is the basic unit of the nervous system. Nerve cells consist of three main parts. Each cell has a round cell body that contains the nucleus. Small extensions radiate from the cell body. These are dendrites. The third part is the axon. The axon is a longer, thicker extension that carries signals to other nerve cells or to muscle.

Glia are cells that cover nerves and the entire nervous system. They help with signal transmission along the axon's route. In addition, glia play a role in blood vessel dilation. Their purpose here is to dilate vessels when the neuron needs more oxygen. An increased blood flow in the dilated vessel delivers more oxygen faster than regular flow.

The Circulatory System

The circulatory system has two main functions. First, it helps parts of the body communicate with other parts. It does this not by sending direct signals as in the nervous system, but by using cells and substances that move through its vessels. Second, the circulatory system allows the body to exchange materials with the environment. This system carries wastes away from tissues. Specific organs will then remove those wastes. For example, the lungs expel carbon dioxide and the kidneys excrete nitrogen compounds. Meanwhile, the circulatory system carries the oxygen that you draw in from the environment with every breath. All other tissues can keep working by feeding this oxygen into their energy pathways.

The animal world has two major types of systems that carry out these two functions. One is found in primitive animals. It is called an open circulatory system. Higher animals possess a closed circulatory system.

Open and Closed Circulatory Systems

The most ancient animals on Earth dispense with a circulatory system. Prokaryotes and the simple eukaryotes get oxygen (or carbon dioxide) by diffusion: the gas oozes across the cell membrane. No circulation is required.

As organisms get more complex, the task of getting oxygen to all tissues becomes difficult. Some invertebrates, such as jellyfish and flatworms, solve the problem with an inner cavity filled with fluid. This allows oxygen to diffuse from the fluid into the cells.

Animals more complex than jellyfish, such as insects and mollusks, have an open circulatory system. In this system, a heart-like tube pumps fluid into many sinuses that infiltrate the tissue. Gas exchange then occurs where sinus and tissue meet.

Higher animals rely on a closed circulatory system. In this scheme, a heart pumps blood into vessels that branch into highways and avenues to reach all tissues.

Biology Tidbits

The Heart

The human heart has four chambers separated by four valves and is situated at the hub of major arteries and veins. Cardiac muscle contracts and relaxes in a set rhythm. Two chambers (left and right ventricles) pump harder than the other two (left and right atria [singular: atrium]) because the ventricles must pump oxygen-containing blood throughout the body. When the heart relaxes, it again fills with blood.

The volume of blood pumped per minute is the cardiac output. Two factors affect cardiac output. The first is heart rate or the rate of contractions. The second is stroke volume, which is the amount of blood pumped in each contraction. A good cardiac output improves oxygen delivery to all the tissues that need it.

Blood Vessels

Blood vessels are pliable, flexible tubes that carry oxygen-containing blood toward hungry tissues and carbon dioxide away from working tissues.

The three main kinds of vessels are arteries, veins, and capillaries. Together, these vessels travel 62,000 miles in a single human body! Arteries carry blood away from the heart toward tissues; veins carry blood from tissues toward the heart. The main interface between the circulatory system and tissues occurs in a third type of vessel called the capillaries. These vessels have very small diameters that sometimes allow red blood cells to flow only in single file. Capillaries that form complicated highway interchanges made of hundreds of vessels to feed tissue are called capillary beds.

Do not confuse arteries and veins based on whether they handle oxygen-containing blood or oxygen-depleted blood. We associate arteries with oxygenated blood, but some carry depleted blood. The same is true for veins. Arteries and veins can only be distinguished by the direction they flow, either toward the heart or away from it.

Blood

The term "lifeblood" says a lot about the importance of blood to us. Blood's job in removing wastes from tired tissue keeps us alive every bit as much as blood that delivers oxygen and nutrients. Blood also carries components of the immune system that fight infection, hormones that control specific organs, and particles that clot blood to keep injuries from making you bleed to death.

Blood is mostly water. This water portion, called plasma, carries nutrients, antibodies, clotting factors, hormones, gases, and wastes.

The cellular component contains eight main cell types:

➤ Erythrocytes or red blood cells.

➤ Leukocytes or white blood cells.

➤ Platelets, which are cell fragments used in clotting.

➤ Lymphocytes, which are cells of the lymph system.

➤ Cells of the immune system: basophils, eosinophils, neutrophils, and monocytes.

Plasma makes up about 55 percent of blood and cells make up the rest. The blood that contains both plasma and cells is called whole blood. When a blood sample is treated to remove cells and also the clotting factors in plasma, the resulting liquid is serum.

CHAPTER 23

The Animal Body: The Endocrine, Immune, and Sensory Systems

In This Chapter

➤ The main features of the endocrine, immune, and sensory systems

➤ The endocrine glands and what they do

➤ The cells and antibodies that give us immunity

➤ An overview of the sensory system and sense receptors

In this chapter we continue the theme of what makes higher organisms "higher." The main reason higher organisms are complicated yet also have special abilities comes from communication. Organisms such as humans have sophisticated processes for getting organs and body systems to work together.

This chapter covers three more systems that play a major role in communication with both the internal workings of the body and the external world: the endocrine, immune, and sensory systems.

Before plunging in, let's review the main job of each of these systems:

➤ Endocrine—The system of hormones, substances that are made in one part of the body to regulate an organ in another part of the body.

➤ Immune—The system that fights infection.

➤ Sensory—The system that enables you to perceive and interpret the world around you.

The Endocrine System

This system performs specific communications inside your body, and they are more complex than nerves and blood. Animals and plants both have endocrine systems. We will focus mainly on the animal system here.

The endocrine system consists of all the body's hormone-secreting tissues and organs. If we are speaking about a specific organ, such as the pancreas, then we refer to it as an endocrine gland.

A portion of the endocrine system overlaps with the nervous system, sometimes making these two systems difficult to tell apart. For instance, nerve cells called neurosecretory cells secrete hormones into the blood just like endocrine glands.

Endocrine Glands

Humans have eight major endocrine glands. Their activities are regulated by a variety of factors. For example, the pancreas responds to blood glucose levels to regulate insulin. The adrenal glands, by contrast, are controlled by the nervous system. Some endocrine glands are controlled by the hormones released by yet other endocrine glands. For example, the reproductive glands are controlled by hormones released from the pituitary gland. Yes, the endocrine system is a complex organization.

The endocrine glands of the human body are:

> ➤ Adrenal glands—Release epinephrine and norepinephrine, which affect blood glucose and other aspects of metabolism; release corticoids that raise blood glucose and affect kidney function.

> ➤ Hypothalamus—Releases various hormones that influence the pituitary gland.

> ➤ Pancreas—Releases insulin to lower blood glucose and glucagon to raise blood glucose.

> ➤ Parathyroid gland—Releases parathyroid hormone to raise blood calcium.

> ➤ Pineal gland—Releases melatonin, which helps body rhythms and responses to light/dark cycles.

> ➤ Pituitary gland—Releases a variety of hormones with functions such as bone growth, milk production, mammary gland contraction, water retention by kidneys, reproductive cells, and stimulating other endocrine glands.

> ➤ Reproductive glands (gonads)—The testes produce androgens; the ovaries produce estrogens and progesterone.

> ➤ Thyroid gland—Releases calcitonin to regulate blood calcium and other hormones.

Biology Tidbits

The Pituitary Gland

The pituitary is the most multipurpose of all adrenal glands. This gland is responsible for the body's growth by releasing growth hormone that affects various body tissues.

Other body functions the pituitary stimulates are: milk secretion by prolactin; sperm and egg production by follicle-stimulating hormone; mammary gland and the uterus contraction by the hormone oxytocin; and kidney water retention by antidiuretic hormone. The pituitary gland also affects other endocrine glands. It regulates the testes and ovaries by luteinizing hormone, controls the thyroid gland by thyroid-stimulating hormone, and stimulates secretions by the adrenal glands through adrenocorticotropic hormone.

Hormones

Hormones come in a variety of styles. They can be made of protein, *peptide*, glycoprotein (a protein with a sugar attached), or steroid.

A variety of signals picked up by the nervous system and other tissue tells an endocrine gland to "Wake up!" This communication involves three steps:

➤ Reception—Binding of a stimulating substance to endocrine gland cells.

➤ Signal transduction—Triggering of a series of steps inside the endocrine cell to set up a response.

➤ Response—A change in behavior by the endocrine cell.

In a general sense, hormones devote themselves to maintaining the body's homeostasis. Every adjustment they make to body systems, organs, or cells relates to keeping you in balance.

Pheromones

Some vertebrates and plants communicate with hormone-like chemicals called pheromones. Pheromones differ from hormones by being made specifically for communication between individuals of the same species without using conventional communication, such as calls, howls, barks, etc. An individual releases a pheromone into the environment and another individual absorbs it and responds to it.

Pheromones can be released into the air to be detected by others, but in some species they are excreted onto the body's exterior. A member of the same species detects the pheromone by touching the secretor's body.

There are two main types of pheromones with their own purpose. Sex pheromones affect breeding behavior, including the power to attract a prospective mate. Aggregation pheromones serve to make members of a species move together to form a colony or community. Hundreds of species are known to secrete aggregation pheromones. Insects have received the most study, showing that aggregation pheromones are helpful in protecting species against predators, in sending signals related to mating, for locating other insects, in communicating information about food sources, and for laying down trails.

Plant pheromones serve similar roles in protecting plant species. Some plants release a pheromone when a herbivore takes a bite out of them. This alarm pheromone—insects emit similar alarms when attacked—alert other plants to strengthen their defenses. Plant pheromones also participate in several interactions with insect pheromones. In some cases, the plant chemical repels predatory insects. In other cases, the plant uses a pheromone to attract pollinating insects.

The Immune System

The immune system is an internal protection system. It does two things to help you have a long life: (1) it fights unexpected infections, and (2) it prepares to defend you against repeated infections.

The immune system is based on cells and cell secretions. Their primary concern is to guard your body against any substance that they perceive as being foreign. Any microbe, pollen, protein, transplanted organ, or prosthetic device that the immune system does not recognize will likely draw a vigorous and coordinated attack. Biologists call this talent of the immune system the ability to tell "self from nonself."

In most cases, you depend on your immune system to swiftly and efficiently eliminate any foreign matter that gets inside your body. The immune system does this daily without your knowledge. Its efficiency comes from a multipronged approach to stopping foreign invasion.

The immune system is a body system that gives you immunity. Immunity is the ability to fight infection. (For simplicity, we will think of infection by microbes or a protein. We will not cover here the immune response of your body to a transplanted organ or prosthesis.)

Components of the Immune System

Immunity may be the most specialized activity carried out by your body. When all these components work together correctly, they can spot a microscopic invader. They furthermore know who the invader is and respond against the specific microbe and only that microbe! But at the very same time, the immune system also has components to provide general nonspecific protection against any microbe.

The lymph system acts as part of the immune system. Lymph is a clear fluid that moves in vessels often running parallel to blood vessels. The system's main function is to supply lymphocytes to the body. These cells wander through vessels and into tissue, always on the lookout for an infectious agent. The lymphocytes can act against invaders they find, but the big guns of immunity come later from more specific immune system soldiers.

The components of the immune system are:

➤ Lymph nodes and vessels—Nodes act as congregation points for lymphocytes and foreign matter filtered from the blood. The vessels enable lymphocytes to travel throughout the body.

➤ Thymus gland—Place where certain lymphocytes mature.

➤ Spleen—A giant version of a lymph node where foreign matter is put in contact with immune system cells.

➤ Peyer's patches—Thickened sections of the lining of the intestines that look out for microbes intent on infecting you from the digestive tract.

➤ Bone marrow—Production site of blood cells, including lymphocytes.

Lymphocytes never work alone. The immune system consists of an exquisite network of systems working together and independently. If an infection gets past one defense, another defense steps up to continue the battle. Biologists say that the immune system has redundancy and duality. No other body system can claim both to the degree possessed by the immune system.

Biology Vocabulary

Redundancy and Duality

Redundancy in the immune system means that more than one component does the same job. This seems an inefficient way for the body to work but, on the contrary, the redundancy gives an extra layer of protection against infection. A microbe that has learned to get around one defense might be thwarted by another defense that works in a different manner.

Duality is a hallmark of immunity. In duality, the immune system has two arms that work independently to stop infection. Like redundancy, this scheme gives the body extra insurance against the wiliest of invaders.

Cells of the Immune System

Two types of lymphocytes patrol your body. The first, called B cells, produces antibodies. The second type, T cells, attacks foreigners by themselves but also sends out an SOS to other cells of the immune system that receive the distress call and rush in to help.

In addition to lymphocytes, another important cell of the immune system is called a phagocyte. Several different types of this white blood cell exist. They all perform phagocytosis, which is the act of enveloping a particle and then digesting it. This ensures the invader's destruction. The body has six different types of phagocytes.

Antibodies

Antibodies are proteins designed by the body to latch onto specific invaders. Shaped like the letter "Y," the tip of each antibody contains a surface that fits together perfectly with the structures on the outside of the invader. When antibodies and invaders meet in the blood or in tissue, biologists call the invader an antigen. The antibody-antigen connection represents one of the most specific chemical unions in nature.

The body produces five different types of antibodies that all incorporate the basic Y shape. Antibodies have a modest effect in fighting an infection by a microbe that has never before entered your body. But after that infection has been conquered, B cells store information about the nefarious interloper. Should the same microbe infect you again, the cells launch a concentrated attack against it. Antibodies seek the infectious agent, and because antibodies are very specific, they hang onto the microbe until phagocytes show up to kill it.

Antibodies are so crucial in infection fighting, medicine uses vaccines to help the body make more antibodies against specific pathogens. Thus, immunity comes from two sources: natural and acquired.

Natural Immunity

Natural immunity occurs in everyone with a healthy immune system in two ways:

> ➤ The antibodies we receive from our mother before birth.

> ➤ The antibodies that develop in response to having had an infection.

Acquired Immunity

We can help stop the spread of specific infectious diseases by getting vaccinated against the disease. This gives us acquired immunity. Two types of acquired immunity are available to us, and these types can be thought of as active or passive approaches to becoming immune to an antigen:

➤ Active acquired immunity occurs when you receive a vaccine containing an antigen. When the antigen enters your body, it prompts the immune system you make antibodies against it. The body will then make the same antibodies every time it confronts the same pathogen.

➤ Passive acquired immunity occurs when the vaccine contains antibodies. Vaccination puts these fully formed antibodies into your body and results in instant immunity.

The Sensory System

The sensory system works closely with your nervous system to enable you to sense conditions of your surroundings. The entire system's ability rests on numerous sensory receptors of your body. These receptors are mainly in nerves and skin. They produce sensations when activated by a stimulus. A stimulus might be a flash of light, a clap of thunder, or a blow to your ribs. The correct sensory cells must interpret the corresponding stimulus or you remain blissfully unaware of your environment. No matter how brightly the sun shines, your ears really don't care! Sensations therefore start when a stimulus activates the appropriate receptor.

A sensation is an action potential that swiftly travels along the nerves to the brain. Once the brain gets this message, it interprets what it has learned. This interpretation is called perception.

Types of Sensory Receptors

The body has five main types of sensory receptors to handle almost any condition you become exposed to in the environment:

➤ Chemoreceptors—These receptors are vital for survival in simple organisms because they relate directly to the organism's search for food and escape from harmful substances. In humans, chemoreceptors detect blood levels of glucose, oxygen, and carbon dioxide as well as serve in the senses of taste and smell.

➤ Electromagnetic receptors—These receptors detect light, electricity, and magnetism.

➤ Mechanoreceptors—These are the receptors of touch and are used for sensing pressure, stretching, motion, and sound. They are called mechanoreceptors because these are forms of mechanical energy.

➤ Pain receptors—These are neurons that probably need no further explanation. Pain sensations are an important survival aide because pain causes an animal to flee danger or an attack. You may mull that over the next time you're sitting in the dentist's chair.

➤ Thermoreceptors—These receptors sense heat and cold as well as heat-cold intensity.

Biology Vocabulary

Mechanical Energy

This is the energy of movement. Mechanical energy is the sum of two types of movement-associated energy. First, kinetic energy is the energy of a moving object, such as a ball rolling down a sloping driveway. The second type, potential energy, is stored in the ball when it sits at the top of the driveway before it starts to roll downhill.

Components of the Sensory System

The main places where you sense various characteristics of your environment are your skin, eyes, ears, nose, and taste buds.

You also sense and interpret less obvious stimuli compared with a poke in the ribs, a sunset, a brass band, a rose, or a Philadelphia cheesesteak! The sensory system helps you interpret and adjust to the sensations of gravity and equilibrium. Mechanoreceptors participate in the interpretation of gravity's effects on the body, especially as you move.

Most of our detection of equilibrium, or body position, occurs in the ears. The same receptors that detect sound waves also detect changes in equilibrium. Different body angles affect the tiny hairs attached to sensory receptors of the inner ear. The brain interprets the direction of movement of the hairs and uses the information to determine body position. Invertebrates have very refined systems for assessing gravity, and insects, fish, and birds also have sophisticated equilibrium-sensing systems.

CHAPTER 24

The Basics of Animal Reproduction

In This Chapter

➤ Animal reproductive cycles and asexual and sexual reproduction

➤ Components of the female and male reproductive tracts

➤ Breeding and the fertilization process

In this chapter the discussion turns to animal reproduction. This chapter covers the main parts of the female and male reproductive systems and some important differences between vertebrate and invertebrate reproduction.

Most higher multicellular animals use sexual reproduction. Sexual reproduction hinges on the formation of two haploid reproductive cells that merge in fertilization to form a new diploid cell. The diploid cell contains a mixture of genetic material from two different donors who contribute an egg or a sperm cell.

By contrast, asexual reproduction does not require two different individuals to supply reproductive cells. Offspring get genes from a parent cell by way of mitosis. The reproduction requires no union of egg and sperm cell. Asexual reproduction is the main form of replicating in prokaryotes. Fungi and plants have asexual portions in their reproductive life cycle, too.

Although higher animals use sexual reproduction exclusively, the animal kingdom manages to make room for asexual reproduction. Many invertebrates use asexual means of propagating a new generation.

This chapter covers the highlights of asexual reproduction in animals, and then discusses the components of sexual reproduction.

Reproductive Cycles

Most animals have a reproductive cycle. This cycle is divided into periods of breeding, raising offspring, and nonbreeding. The breeding phase ensures a new generation will arise. In this way, a species survives. The nonbreeding phase helps animals rest, rebuild their strength, and conserve resources. In nature, nonbreeding seasons usually correspond with seasons when the food supply is scarce. In temperate zones, this means winter. Breeding is timed so that when offspring arrive, the climate and food supply favor the newborns. Thus, most species produce their young in the spring when plant life is richer and also when other animals are birthing. Carnivores take advantage of burgeoning populations of prey animals to feed their own young.

Reproductive cycles occur even in regions where seasonal changes are small or nonexistent. This tells us two things. First, the resting phase of the reproductive cycle is important. Second, factors in addition to environment must be controlling the reproductive cycle. Hormones play the role of those invisible factors that drive the reproductive cycle.

Asexual Reproduction in Animals

In animals, asexual reproduction is confined to invertebrates. They perform three main types, depending on the species:

➤ Fission—Splitting into two almost equal halves. Sea anemones use this method of reproduction.

➤ Budding—A small individual, a "bud," arises out of a parent, breaks off, and then grows to adult size. Tunicates are among the animals that use this method.

➤ Fragmentation—A parent breaks into several smaller pieces, and then the pieces grow until they regain adult size. Sponges, tunicates, and a few other primitive animals use this method.

Despite what you may be thinking about asexual reproduction, it offers advantages to animals. Animals using this type of reproduction can live in isolation without fear that they will never find a mate. Asexual reproduction also enables an animal to produce numerous offspring in a short time, another survival tactic. Finally, asexual reproduction causes fewer errors in the offspring's genotype compared with the more random experience of sexual reproduction and fertilization.

Fertilization

Fertilization is the meeting of an egg and sperm cell of the same species to produce a fertilized egg. The animal world uses two types of fertilization:

➤ External—The female releases eggs into a watery environment and a male comes along to release sperm that fertilizes them.

➤ Internal—The male deposits sperm cells in or near the female reproductive tract. Fertilization occurs inside the tract.

Most aquatic species use external fertilization because the water provides a suitable place for egg and sperm to combine. Often, these species use cues from environment, such as water temperature or amount of daylight, to release eggs and sperm cells. Some species rely on pheromones to tell them when the breeding part of their cycle is about to begin.

The animals on dry land had to evolve a reproductive mechanism that kept eggs and sperm cells from drying out before fertilization could be completed. Internal fertilization came about as a way to protect the valuable reproductive cells and ensure breeding success.

Parts of the Female Reproductive System

The human female reproductive system consists of five components and one auxiliary part:

➤ Ovaries—Place where the eggs develop. Humans have two ovaries.

➤ Oviducts—Also called fallopian tubes, these ducts catch an egg as it is released from an ovary. Since the ovaries and oviducts do not directly attach, the ducts have a funnel-shaped opening to help catch the egg.

➤ Uterus—This is the womb where a fertilized egg becomes a growing embryo and develops into a fetus.

Biology Vocabulary

Mammary Glands

Mammary glands occur in both sexes in animals, but only the female has enlarged, functional glands. In females, epithelial tissue secretes milk. The milk drains into a network of ducts leading to the nipple. This is the place where a nursing baby suckles, or receives milk from its mother. Around the ducts, the mammary glands are mainly fatty tissue, and in nonlactating females, the fatty tissue makes up almost all of the gland's tissue.

Estrogen maintains the female mammary gland. Because males have very little of this hormone, their mammary glands do not develop fatty tissue or a duct system.

➤ Vagina—Canal in the reproductive tract for the deposit of the male's sperm cells.

➤ Vulva—The collective term for the female's external genitalia and the opening of the reproductive tract to the outside of the body.

➤ Mammary glands—These are the auxiliary components. These are enlarged endocrine glands that produce and secrete milk in the female after giving birth.

Parts of the Male Reproductive System

Unlike females, which have internal reproductive organs, the male's reproductive system is located partly inside and partly outside the body.

The external parts are the scrotum and penis:

➤ Scrotum and testis—A fold of the body wall that holds the testes, the site of sperm cell production.

➤ Epididymis—Tubes connected to the testes and in which new sperm cells mature and become motile.

➤ Penis—Part of the duct system that deposits sperm cells in the female reproductive tract.

A male's external genitalia must be located outside the body because sperm cells cannot mature at body temperature. Most mammals therefore use this method of holding the testes at a slightly cooler temperature to help sperm production.

The internal parts consist of organs that help with ejaculation, propelling the sperm cells into the female reproductive tract. Accessory glands also make up the male internal reproductive system:

➤ Vas deferens—A tube that delivers the sperm cells from the epididymis to the penis during ejaculation.

➤ Ejaculatory duct and urethra—The tubes that carry the sperm cells through the penis.

➤ Seminal vesicles—An accessory gland that contributes most of the fluid that carry sperm cells during ejaculation. The sperm-containing fluid is called semen.

➤ Prostate gland—An accessory gland that contributes to the fluid portion of semen.

➤ Bulbourethral glands—Accessory glands that produce a small amount of mucus just before ejaculation. This mucus neutralizes any acidic urine that might be in the urethra; the acidic urine would otherwise kill most of the sperm cells.

Biology Vocabulary

Semen

Semen consists of three components: sperm cells and fluid from two different glands. About 60 percent of the fluid in semen comes from the seminal vesicles. These glands produce fluid that is thickened with mucus and carries sugar, which the sperm cells use for energy. The remaining 40 percent of the fluid comes from the prostate gland. Its fluid is thin and milky.

The fluids of semen also contain smaller quantities of coagulating enzyme, ascorbic acid, citrate, and prostaglandins. These components help regulate the conditions for keeping sperm cells alive and enhancing the environment inside the female reproductive tract.

In humans, semen volume is about two to five milliliters per ejaculation, and each milliliter contains about 100 million sperm cells.

Breeding

When a male deposits sperm cells in the female reproductive tract, conditions must be right for getting the sperm to the egg located in the oviduct. Semen is alkaline and neutralizes acidic conditions of the vagina. Prostaglandins in semen help thin the mucus and induce contractions in the uterine wall, which help push the sperm cells up the uterus. The coagulating enzymes also bunch up sperm cells to make more of them move with each uterine contraction. Later anti-coagulating enzyme breaks apart the clumps.

Fertilization of the egg occurs in the oviduct. As with most complex events in biology, fertilization consists of steps that each must unfold correctly for the entire process to be successful:

➤ Sperm and egg contact—A sperm cell contacts the outer coat of the egg. This triggers the release of enzymes from the tip of the sperm cell.

➤ Acrosomal reaction—The sperm's enzymes eat a hole in the egg's coating and the sperm produces a protrusion that connects with the egg. This protrusion is called an acrosomal process.

➤ Fusion—The sperm cell membrane fuses with the egg membrane.

➤ Entry—The sperm's nucleus enters the egg's cytoplasm.

➤ Cortical reaction—The fusion step simultaneously causes a release of calcium ions from the egg's membrane. This step and other adjustments in the egg clip off the sperm cell's tail and produce a strong envelope around the fertilized egg.

The cortical reaction mentioned above contains a series of steps that complete the fertilization process. Its two main purposes are to accept the sperm cell nucleus into the egg and to change the egg infrastructure to protect what is now a newly formed zygote. The calcium released in this step occurs as a wave that spreads across the egg in about 30 seconds. The calcium ions ($Ca2^+$) depolarize the egg during this period and is part of the egg's activation.

PART SEVEN

Plants

CHAPTER 25

The Plant Body and Plant Growth

In This Chapter

➤ Ways of thinking about plant diversity

➤ Early plant life and today's plant ancestors

➤ The main parts of a vascular plant

➤ The role of roots, stems, and leaves

In this chapter we will explore the diverse world of plants. By "plants," I mean both plants of gardens and meadows as well as tall woody trees. In biology, you only get three choices: microbes, animals, and plants!

Plant Diversity

Plant fossils date to about 475 million years ago and scientists believe that the first green plants took up residence on land about 500 million years ago. Prior to that time, plants lived in water and had evolved from the early cyanobacteria and green algae.

Once on land, plants diverged in their development into hundreds of thousands of species. Adaptations such as leaves enable them to catch sunlight from different angles. Roots attached them to the earth and spread belowground to find nutrients. On early Earth, few plant-eating animals existed, so plants could flourish and diversify.

Modern plant diversity remains enormous even though dozens of species go extinct every month. The vastness of plant diversity enables us to classify plants in many different ways:

➤ Based on evolutionary history.

➤ Aquatic versus terrestrial.

➤ Vascular versus nonvascular.

➤ Soft plants versus woody plants.

➤ Seed-producing versus non-seed-producing.

➤ Flowering versus non-flowering.

➤ Based on climate and geographic habitat.

In each of these categories exists additional sub-categories. For instance, classifying plants by climate and geographic region leads to varieties of tropical, sub-tropical, temperate, boreal, and desert regions, to name just a few groups.

These chapters will cover the land plants and will focus on two ways of classifying plants. The first classification method stressed in this chapter covers the vascular versus nonvascular plants. Vascular plants contain vessels that carry water and other substances throughout the plant body. Nonvascular plants lack these vessels and are more primitive in evolutionary terms that their vascular cousins.

A second major way to group plants relates to their evolution. The nonvascular plants arrived here first, then the vascular plants. Of the vascular plants, you will soon learn two additional groups: seed plants and flowering plants.

Early Plant Life

All plants evolved from ancestors that developed two cell types: gametophytes and sporophytes. Both types of cells participate in the plant's life cycle; they differ by the number of chromosome replicates they carry.

Today's plants that most closely resemble ancient land plants are called bryophytes; these are the nonvascular plants. Bryophytes include three main groups:

➤ Liverworts—Flat, low-lying plants with leaflike gametophytes; common in tropical and subtropical regions.

➤ Hornworts—Plants with long vertical sporophytes.

➤ Mosses—Plants characterized by gametophyte growth in mainly a vertical direction and long sporophytes of various colors.

The bryophytes have a distinctive life cycle that features two stages. Bryophytes undergo one stage in which the plant has only one set of chromosomes (the haploid stage) and a second stage whereby the plant possesses a full double set of chromosomes (the diploid stage).

Every haploid stage occurs after the diploid cells undergo meiosis. Put another way, diploid always leads to haploid. After two haploid cells merge by fertilization, a diploid cell called a zygote always results. That's why biologists call the life cycle of bryophytes an alteration of generations.

Bryophytes are also distinctive in the plant kingdom because both the haploid form and the diploid form are multicellular plants. In animals and in higher plants, this is not the case. In higher organisms the haploid reproductive cell is but a single cell. In brypohytes, the haploid multicellular plant is called the gametophyte and the diploid multicellular plant is called the sporophyte. When we get to higher plants in the next chapter, you will see a a much greater difference between these two forms.

Mosses

Mosses grow worldwide and thrive in bleak conditions found on mountaintops, deserts, and the *tundra*. Because mosses overcome rugged conditions, they are among the first plants to recolonize land that has been destroyed by fire. Generations of mosses slowly add nutrients back to the land and improve conditions for the advent of vascular plants. This slow progression of plants from simple species to increasingly complex species is called succession.

Biology Vocabulary

Gametophytes and Sporophytes

In plants that undergo an alteration of generations, such as the primitive bryophytes, gametophytes and sporophytes both take the form of multicellular plants. In more advanced plants, by contrast, the gametophyte has evolved to a single cell. A gametophyte cell holds one set of chromosomes or a haploid set. A sporophyte cell holds both sets of chromosomes or a diploid set.

Peat moss represents a special type of moss growth. This wetland moss forms deposits of organic matter known as peat after generations of dead plants accumulate in the wetland's sediments. The deposits, called peat bogs, serve as a source of fuel. Peat is burned in many parts of the world for heating and cooking.

Ferns

Ferns occupy a bridge in the evolution of vascular plants from nonvascular plants. Ferns are non-seed-producing vascular plants. Like all other seedless plants, ferns reproduce by producing a sperm cell that must swim through water to reach an egg to fertilize. Not surprisingly, these types of plants live mostly in damp environments.

The fern that we are most familiar with is the sporophyte. This leafy form of the plant is a precursor to the vascular leaves of later plants. The underside of each fern leaf holds small dots called sori. Each sorus is a cluster of sporangia, which are sacs that contain spores. When released, spores travel through the environment until they find a suitable bit of land to settle. A new gametophyte emerges from the spore and begins growing into a multicellular gametophyte plant, which must be fertilized to produce a new sporophyte. The young fern supplies both the sperm cells and the eggs needed for this fertilization.

Vascular Plants

Because vascular plants are so diverse, botanists put them in groups called clades. All the vascular plants themselves can be considered a clade that accounts for more than 90 percent of all the Earth's plants!

As mentioned earlier, these plants belong to two groups, each containing several thousand species:

> ➤ Seed plants called gymnosperms, which bear "naked" seeds that are not enclosed by a protective covering.

> ➤ Flowering plants called angiosperms, which produce seeds that are enclosed in a protective ovary.

Biology Tidbits

The Earth's First Land Plants

Modern land plants evolved from green algae. A critical step in this evolution occurred when algae moved from living exclusively in water to life on land. Certain algae called charophycean algae possessed a layer of tough material in their outer coat that kept the cell from drying out if it was out of the water for prolonged periods. This simple adaptation led to plant cells more suited for life on land than their aquatic cousins. A photosynthetic cell that could live permanently in places bathed in bright sunlight and immersed in an atmosphere with plenty of carbon dioxide enjoyed an advantage over water-bound plants.

Green land plants developed about 500 million years ago. Their diversity is enormous, which attests to the advantages of living on land. Of the plant ancestors, only sea grasses migrated back into water environments.

Anatomy of a Plant

Green plants consist of four main parts. From bottom to top they are the roots, stem, leaves, and flowers (in flowering plants only). Each plant part has specific jobs in keeping the plant alive and growing.

Roots

Roots are the underground part of the plant. They stabilize the plant body by anchoring it and also absorb nutrients and water from the soil.

The main root that grows in a downward direction is the primary root. Lateral roots branch off the primary root and grow horizontally. Root growth occurs at the root tip, which is a region of rapid cell growth.

The secondary roots that branch laterally from larger roots are covered with tiny hair-like projections, aptly named root hairs. These projections greatly increase the root system's surface area for taking in nutrients. The root hairs also come in contact with a distinctive soil region that is in closest contact with the roots, called the rhizosphere. The rhizosphere represents an area of active nutrient and gas exchange between the plant and soil microbes.

Biology Vocabulary

The Rhizosphere

The rhizosphere is the region of soil that lies right next to roots. This region extends only five millimeters (two-tenths of an inch) from the root surface. The rhizosphere's levels of carbon dioxide, oxygen, water, and acids differ from that in any other part of soil.

A variety of microbes also exist in the rhizosphere; some of them are found in few other places. For example, specialized bacteria perform nitrogen fixation. In this process, the bacteria pull nitrogen gas from the air and convert it to a form that the root can absorb and the plant can use to make proteins. Certain fungi also play a distinctive role in the rhizosphere. These fungi cling tightly to the root and form extensions that meander into the soil. These extensions draw in extra nutrients for the plant and so increase the plant's chance of survival.

Stems

Stems give the plant its aboveground structure from which leaves branch. Stems also strengthen the plant body due to tough fibers that form a net-like matrix in combination with polysaccharides and proteins. The main plant fibers of the stem are cellulose and lignin.

Cellulose contains many repeating units of glucose, similar to how starch is composed. Cellulose differs from starch in how the glucoses chemically bind to each other. These bonds give cellulose its strength. Lignin provides even more rigidity than cellulose. Lignin contains a variety of sugars and other organic compounds that make it nearly indestructible.

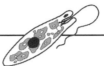

Inside a Cell

Plant Cell Walls

Plants contain a rigid cell wall that envelops the cell contents and which animal cells lack. The thick plant cell wall holds the cell's shape, and collectively, the cell walls keep the plant upright. The cell wall's strength comes from the fiber cellulose, which together with polysaccharides and proteins form a fiberglass-like matrix. For the plant's added strength, cell walls of adjacent cells connect with each other with a glue-like substance that hardens as the plant matures.

When you eat vegetables and fruit, you add fiber to your diet. This benefit comes directly from the plant cell wall!

In addition to giving the plant its main support structure, the stem of vascular plants also holds the vessels called xylem and phloem. These vessels play the essential role of transporting water and nutrients to hungry plant cells and sugar from the cells that perform photosynthesis.

Some plants diversified during their evolution to form modified versions of the stem. These modifications are familiar to most of us:

> ➤ Bulbs—Underground enlarged sections of the stem, usually containing many layers, such as seen in onions.

> ➤ Rhizomes—An enlarged stem that grows horizontally just beneath the soil surface, such as ginger root.

➤ Tubers—Enlargements at the end of a rhizome. The most familiar tubers are potatoes, sweet potatoes, and yams.

➤ Stolons—Stems that grow horizontally along the soil surface as in strawberry plants.

The straight sections of the stem are called shoots, as they are in elongated sections of roots. The points on the stem onto which leaves attach are called the nodes.

Leaves

Leaves keep plants alive. They serve as the site of photosynthesis, carbon dioxide uptake and glucose manufacture, energy storage, gas exchange between the plant and atmosphere, and help in water transport. A plant's leaf cells contain thousands of chloroplasts for converting the Sun's energy into chemical energy via photosynthesis.

The glucose produced by photosynthesis may be stored in the leaf, usually as polysaccharide or cellulose, or shipped to other parts of the plant to make proteins, fats, and other organic compounds.

Leaves also protect the seeds of some plants, catch rainwater, act as the site for evaporation of transpired water, and provide a place for gas exchange with the atmosphere. Carbon dioxide enters the plant and oxygen exits through small leaf pores called stomata.

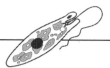

Inside a Cell

Xylem and Phloem

Xylem and phloem are the pipes of vascular plants. Xylem conducts most of the plant's needed water and minerals, often in an upward direction from the roots. The phloem carries the sugar glucose and other organic nutrients in a general downward direction throughout the plant.

Biology Vocabulary

Transpiration

Transpiration is the loss of water from a plant by evaporation from the leaves. Water comes into a plant mainly through the roots. The plant transports it upward in xylem and sends it out to the leaves where the water is needed in photosynthesis. Excess water returns to the atmosphere. Transpiration is in this way an important piece of the Earth's water cycle.

Botanists and home gardeners have long used leaves as a way to identify plants. With a little practice, almost anyone can group plants by leaf shape and arrangements.

Leaves are either simple or compound. Simple leaves consist of a single leaf blade attached to the stem's node. Compound leaves contain a central stalk (called a petiole) connected to the node, and from this stalk branch several small leaves. A double compound leaf consists of small stalks off the petiole, with each small stalk holding several miniature leaves called leaflets.

Compound arrangements include palmate in which several single leaves meet at a central point on the stalk, or pinnate in which many leaves attach at different points along the central stalk.

A leaf's shape also helps in plant identification:

> ➤ Entire—The leaf edge forms a single smooth line.

> ➤ Toothed—The leaf edge contains many sharp points or spikes.

> ➤ Lobed—The leaf edge consists of several rounded protrusions like hills and valleys.

Other than a gorgeous flower, few things say "plant" as well as a leaf.

Biology Tidbits

Wood

Trees and many bushes produce a hard material we know as wood. As a tree ages, many of xylem toward the center of the stem (or trunk) stop transporting water and start to accumulate *resins*. The resins turn this inner column of vessels, called heartwood, hard and darker than the rest of the tree. Heartwood is the material harvested as lumber. The outer layers of xylem continue transporting to keep the tree alive. This functioning collection of vessels makes up sapwood.

The entire tree trunk is a source of wood depending on the tree species. From the innermost region to the outer covering of the trunk, trees consist of: heartwood, sapwood, vascular cambium (primary location of active xylem and phloem), and multiple layers of bark.

CHAPTER 26

 # Seeds and Plant Life Cycles

> ## In This Chapter
>
> ➤ Learn about plant life cycles
> ➤ The importance of seed plants and seeds

In this chapter you become familiar with plant life cycles. Life cycles occur in various complexity all living things, even a simple bacterial cell. In plants, a life cycle can become quite complicated, but all have the same purpose: to produce a new generation of plants that carries its parents' traits in its chromosome. Like animals, plants rely on meiosis and mitosis to accomplish this.

I will begin by describing a plant life cycle. From the simplest mosses to the mightiest pines, the life cycle ensures that a plant's population will continue. Unlike most animals, these life cycles can extend over decades or even centuries.

What Is a Plant Life Cycle?

A plant life cycle is a series of phases through which a plant goes to make different forms of itself. A seed popped loose from a pinecone doesn't resemble a towering Douglas fir, and yet they both belong to the exact same species. Imagine if your life cycle demanded such drastic physical changes!

The hallmark of life cycles of land plants is a process called alteration of generations. In this process, a multicellular plant goes through a phase in which its cells are haploid (containing one set of chromosomes) and another multicellular phase where the cells are diploid (containing the full double set of chromosomes). This differs from animal life cycles in which haploid cells are always single cells of the reproductive system. In plants, the haploid generation always gives rise to a diploid generation, and the diploids always then lead to haploids.

Two different types of cells characterize the alternating generations: gametophytes and sporophytes.

Biology Vocabulary

Gametophytes and Sporophytes

The gametophyte form of land plants contains a single set of chromosomes. In mitosis, the gametophytes produce gametes, either an egg or a sperm, and the two gametes fuse during fertilization to form a new diploid cell called a zygote. The zygote keeps its full set of chromosomes as it undergoes mitosis. This mitosis produces another multicellular form, the sporophyte. Sporophytes contain a full double set of chromosomes. Sporophytes then undergo meiosos to produce haploid spores. Plants can disperse spores into the environment. When the spore settles into soil, it begins to grow into a new multicellular plant. Because this plant contains only one set of chromosomes provided by the spore, it is another gametophyte and ready to continue the plant cycle.

A bryophyte's life cycle includes repeating phases that alternate between gametophytes and sporophytes. These plants use spores and gametes as the haploid form and zygotes as the diploid form. When vascular plants evolved, they became less dependent on the gametophyte form.

When you find ferns growing in the woods, the large fronds you see are part of the fern sporophyte. The woods themselves—the trees tall and short, bushes, and vines—are all sporophytes. In these plants, the gametophytes are usually hidden inside a small part of a cone or flower. Some gametophytes are microscopic.

Smaller sized gametophytes represent a step forward in plant evolution for the following reasons:

➤ The small size allows the gametophyte to remain enclosed in a structure that protects it from environmental stresses, such as drought or excess sunlight.

➤ The small size reduces the nutrient demands of the gametophyte as it draws food from the sporophyte.

➤ The sporophyte plant can package the gametophyte in a coating that protects it during dispersal in the environment.

As plants evolved to make smaller gametophytes, man also developed more prominent seed forms. Seeds further helped plant evolution and survival by enabling plant embryos to withstand harsh environments without dying. Today many varieties of seeds withstand freezing, heating, drying, ultraviolet radiation, and even chemicals. Most seeds stand up to heavy duty physical pounding too.

Before we examine the life cycle of a seed-producing plant, let's take a closer look at what makes seeds special.

Seeds

Seed plants develop female structures called ovules. Many species also produce a male structure called pollen. A seed begins as an ovule that contains some fleshy diploid material (megasporangium) and some haploid material (megaspore). An egg sits nestled inside the megaspore. A thick protective layer called the integument surrounds most of these materials. A feature of the integument is a small opening (micropyle) to the environment. The only way pollen gets to an unfertilized egg is through this micropyle.

When a pollen grain fertilizes the egg, the megaspore enlarges, but the strong integument stays in place around it. The fertilized egg develops into an embryo inside this new seed. The soft remaining haploid tissue surrounds the embryo and serves as its food supply. All the while, the integument turns harder and becomes a rigid seed coat.

A seed is therefore a self-contained, resistant, and durable structure that protects, nurtures, and even feeds the infant plant or tree.

Biology Vocabulary

Pollen

You may know pollen only as the stuff that makes you miserable during allergy season. Pollen is the term for tiny particles calls grains that contain the male gametophytes of seed plants. In other words, pollen grains are male sex cells analogous to an animal's sperm cells.

Seed plants produce pollen grains in an infinite number of shapes. The grains usually measure from 10 to 85 micrometers. The pollen surface contains thousands of spikes and crevices that help keep the particle airborne during dispersal and also help it attach to surfaces when it lands. Botanists examine the distinctive pollen shapes and spike patterns and thus can often identify the plant that made the pollen.

Pollen is a serious *allergen* in many people. Most towns publish a daily pollen count for the air to help allergy sufferers know when it's best to avoid the outdoors. Pollen counts of 0 to 15 mean conditions are healthy for allergy sufferers and asthma patients. Counts in the 16 to 500 range correlate with higher risks. Pollen counts above 500 are considered dangerous for people sensitive to certain types of pollen.

Why Seeds Are Important to Us

The indestructible nature of seeds gives their species obvious advantages in survival. Seeds can hold out in dormant form during poor environmental conditions. When the amounts of water, light, and nutrients and temperature improve, the seed begins to grow into a new plant. This process is called germination. The robustness of seeds helps protect plant diversity, but seeds also impact humanity.

Seeds maintain forests and farm crops, so directly or indirectly they provide us with food, fuel, wood, and some medicines. Humans tend to rely on gymnosperms for fuel and wood products; we use angiosperms for raising crops and providing drugs. For example, six angiosperms—wheat, rice, corn, potatoes, sweet potatoes, and cassava—supply about 80 percent of all the calories consumed by humans. Angiosperms also provide the majority of feeds used for raising meat-producing animals. Seeds furthermore sustain hundreds of species of birds and supplement the diets of many mammals, reptiles, amphibians, and other wildlife.

A small sampling of the human drugs derived from seed plants are:

> ➤ Atropine, used for dilating the eyes comes from extracts of the belladonna plant.

> ➤ Menthol, an ingredient in cough and sore throat medicines, comes from the oil excreted by eucalyptus trees.

> ➤ The drug taxol, used in treating certain cancers, is extracted from bark of Pacific yew.

> ➤ Morphine, a pain reliever, comes from opium poppies.

Society has domesticated plants in a way similar to domesticating animals. Domestication has altered the variety of certain plants because farmers selected the seeds of their favorites. The best flowering plants or best food-producing plants became more prevalent than other types of plants. As a result, some seed-producer species number in the hundreds of thousands and other species number less than 100 worldwide.

Of course, many seed plants proliferated with unintentional help from people and animals. Primitive societies that ate seeds distributed the seeds in their wastes. Domesticated animals, wildlife, and birds did the same thing and continue to do so today. Birds remain one of the most effective ways of sustaining plant populations and dispersing their seeds. Wide seed dispersal helps plants breed with members of their species from distant populations. This mixing of two populations' genes increases the plant's genetic variation, which is important in a species' long-term survival.

Life Cycle of Seed Plants

Now that I have emphasized the remarkable value of seeds, I present here a typical life cycle of a seed-producing gymnosperm. This tree is a conifer, that is, a tree that produces cones. In the pine tree in this example, the mature tree—a sporophyte containing a full set of chromosomes—possesses two types of pine cones:

➤ Ovulate cones contain the female ovules.

➤ Pollen cones contain the male pollen grains.

The life cycle begins when pollen grains escape from the pollen cone and find their way into the ovulate cone. A pollen grain comes in contact with the ovule. This step is called pollination. Because both the female and male cells had undergone meiosis, they carried a haploid version of the chromosome. At fertilization, the two sex cells will merge their complement of DNA and produce the diploid zygote or embryo inside a new seed.

In pine trees, fertilization might occur more than a year after pollination. During the intervening time, the pollen grain eats through the megasporangium to get to the egg. It does this by forming a pollen tube, which worms its way forward. Meanwhile, two or three extra eggs develop. By the time the eggs are ready to be fertilized, the pollen tube has just about reached them.

Fertilization takes place when the pollen tube reaches the egg and the sperm and egg contact each other. They merge their DNA by uniting their respective nuclei. The new cell is now diploid.

After fertilization, the growing embryo begins using up the seed's food supply. The outer seed coat also changes shape so that it will break apart easily as the embryo grows. At germination, a baby seedling pokes through the seed coat.

Although green plants survive by absorbing carbon dioxide, a seed absorbs oxygen. Germination will not occur unless the seed has the right levels of water and oxygen. Temperature also plays a part in when germination starts.

How does a seedling know to grow up toward the soil surface rather than sideways or downward? Gravity, carbon dioxide, and temperature

Biology Vocabulary

Photomorphogenesis

Photomorphogenesis refers to the effects of light on plant structure and growth. "Photo-" refers to light. The term "morpho" stands for morphology, which means the physical features of an organism. Finally, "-genesis" refers to the growth or development of an organism.

may all play a part in guiding the new plant, but light seems to be the most important factor. Seedlings quickly change form into mature-looking plants due to a process called photomorphogenesis. Scientists have studied seedlings that happened to emerge from the seed in darkness. These seedlings slow their root and stem growth until light returns. Then they pick up speed in turning into a mature plant. This temporary pause is probably an adaptation developed in evolution so that plants could get the sunlight so vital to their growth and survival.

CHAPTER 27

 **Flowers, Fruits, and Pollination**

In This Chapter

➤ Parts of a flower and fruit and what they do

➤ How botanists classify flowers

➤ Pollination and the flowering plant life cycle

In this chapter you learn about the flowering plants. These plants avoid spitting their seeds into the vagaries of nature as do gymnosperms. Instead, angiosperms develop a specialized structure for sexual reproduction. You know this structure as a flower. You will also explore the angiosperm structure called the fruit. We know of fruits as a food but much more goes into the development of fruits for a plant's survival.

This chapter also covers pollination, which is a critical step in the angiosperm's life cycle. To understand the life cycle, it helps to know the flower.

Parts of a Flower

A flower is the reproductive organ of an angiosperm. Most flowers contain four basic units: petals, sepals, carpels, and stamens. These parts are described here:

➤ Petals—Leaf-like and brightly colored structures that give flowers their distinctive appearance.

➤ Sepals—Small leaves that sit just beneath the petals and support them. All the sepals collectively make up the calyx. The calyx connects the flower to the stem.

> ➤ Carpels—Female reproductive organ that makes female gametophytes. After fertilization, the carpels act as the sight of seed development. A flower's carpels collectively make up the pistil.

> ➤ Stamens—Male reproductive organ that produces pollen.

Both reproductive organs of flowers consist of smaller parts with specific jobs.

Carpels

A flower's carpel rests at the top of the stem. Carpels consist of three parts. The first is the ovary, which produces the plant's ovules or eggs. The second part, the style, is a tube that sits atop the ovary and transports pollen to the eggs during pollination. The third part, the stigma, is situated at the top of the tube and gives pollen a landing place before pollination. The stigma has a sticky consistency that catches and holds the pollen.

In the life of a flower, petals get noticed. You can't miss their showiness, color, and brightness. Pollinating insects and birds invariably find them. Florists put together flower arrangements based on petals, and they are the reason cut flowers end up on tables at wedding parties and bistros. But the real work in a flower happens at the stigma.

The Stigma

When a stamen releases pollen into the air, pollen must travel distances as short as a half-inch to more than a mile. Long-distance pollen dries out as it goes. Drying weakens most biological things. Pollen grains must furthermore find the correct species of flower to pollinate. What if after being airborne for a day and blowing in the wind over several acres, the pollen finally lands on an appropriate flower only to be washed away in a summer shower? Nature has engineered the stigma to grasp and hang onto pollen as few other things in nature do, even an itchy nose in allergy season!

The stigma has the following features that help increase the chances that fertilization will eventually occur:

> ➤ The stigma contains secretions and has a surface topography that form a strong adhesion to pollen.

> ➤ The stigma adheres stronger to pollen of the same species than pollen from other plant species.

> ➤ Some stigma adhesives help rehydrate the dry pollen grains. This keeps the pollen from becoming dormant and unable to fertilize the ovule.

> ➤ The attraction between pollen and the stigma seems to get stronger as pollination continues.

In addition to the stigma's chemicals and physical features, the outer surface of pollen contributes to adhesion. This outer layer of the pollen grain is called exine.

Stamens

Stamens appear simpler than carpels. They consist of a long filament that originates near the base of the ovary. This filament holds at its apex a small pouch called the anther. The anther produces pollen and stores it until the plant releases the pollen grains.

Biology Tidbits

Flower Seasons

Different plant species produce flowers at different times of the year. The climate of the plant's main geographic habitat determines this schedule. You may be most familiar with plants that flower in spring, summer, and fall, but in places with mild winters, many species produce flowers then too. The time a plant blooms depends on its elevation, geographic location, and amount of sunlight and shade.

Other Parts of a Flower

Two additional parts lie hidden inside the flower. The first, the nectary is a small sac that holds the flower's nectar. This substance attracts insects, which are crucial to pollination by carrying the pollen grains from flower to flower during their nectar hunt.

The second part is the receptacle, which gives the entire flower physical support. It is a bulbous area connecting the calyx and the rest of the flower to the stem. In some flowers, such as strawberry, the receptacle becomes part of the fruit after fertilization.

Biology Vocabulary

Nectar

The composition of nectar varies among plant species. All nectars have in common a high concentration of sugar, mainly sucrose, glucose, and fructose. Many nectars also contain smaller amounts of amino acids and minerals, mainly potassium.

Flowers adjust their nectar secretion depending on their need to be pollinated. For instance, as pollinating insects begin visiting the flower early in pollination season, nectar production starts to increase. After pollination, the flower reabsorbs most of the nectar.

Classifying Flowers

Botanists often classify flowers by plant clades. By this classification method, the flowers of orchids, lilies, poppies, etc., make up individual groups.

Flowers are also grouped according to their anatomy, leading to two main groups:

> ➤ Complete flowers—Contain a pistil, petals, sepals, and stamens.

> ➤ Incomplete flowers—Lack one of the structures of complete flowers listed above.

You can also classify flowers is by the presence or absence of reproductive organs:

> ➤ Perfect flowers—Contain female and male reproductive organs on the same flower.

> ➤ Imperfect flowers—Contain only female or male reproductive organs.

Pollination

The flower provides an elaborate and rather showy place for the main events of a plant's life cycle: pollination and fertilization. Pollination occurs when pollen grains from a seed plant come in contact with the plant's female reproductive organ. Fertilization cannot happen without pollination occurring first.

Some plants self-pollinate. This means that the pollination and fertilization steps take place in the flowers of a single plant. By contrast, cross-pollination happens when pollen from one plant travels to the stigma of another plant of the same species.

Biology Vocabulary

Cross-Pollination

Most plants have flower arrangements that favor cross-pollination over self-pollination. Cross-pollination increases genetic variability because two unrelated plants of the same species mix their DNA when fertilization occurs. This is an advantage to a plant's survival as a species. To reduce the chances of self-pollination, many plants stagger the timing of when their flowers' stamens and carpels mature. This forces the reproductive organs to interact with the opposite sex organs on a different plant.

Insects and Pollination

Wind, rain, insects, and wild animals all carry pollen from flower to flower. Of these, insects are likely the most important factor in moving pollen around and sustaining our plant populations.

Honey bees offer the best example of a relationship that benefits both flower and insect. Honey bees accumulate pollen grains on their body as they attempt to draw nectar out of the flower's nectary. The pollen sticks to the hair on the insect's body like it sticks to the stigma. Honey bees stash nectar in their stomachs for transport back to the hive. In the stomach, enzymes work on the nectar and start converting it to honey, which the bee then deposits into a well in the hive's honeycomb. Pollen simply goes along for the ride! In the process of visiting dozens to hundreds of flowers in a day, a bee moves a lot of pollen around in nature.

Few insects are as important to pollination as honey bees. Some plants cannot survive without the seasonal contribution of these insects in the plant life cycle. Other bees, such as bumble bees and miner bees help in pollination. Wasps and yellow jackets, flies, and butterflies also play minor a role in pollination.

Fruits

A fruit is a mature ovary, although some plants incorporate other parts of the flower into the fruit. The purpose of a fruit is to protect the seed as it matures. Society found other purposes for fruits long ago as a food source. Because the fruit carries nutrients for the developing embryo of the seed, it also serves as a good source of sugars, vitamins, and minerals for people.

After fertilization, a seed develops out of a zygote. In fruit formation, plant hormones induce the ovary wall to grow and thicken around the seed. The fruit often, but not always, forms a fleshy material around a harder, protective seed shell. The hard shell is called a pericarp; you know it as a pit. Thus, fruits such as peaches, pears, and apples contain a fleshy part and a pericarp.

In some fruits, such as grapefruit, the pericarp is rather soft and is surrounded by an equally soft fleshy material. In grapefruit and other citrus fruits, a strong outer rind provides most of the protection to seeds. Grapes and berries offer an example of fruits that start out with a strong pericarp that softens as the fruit matures.

Tomatoes may be most vulnerable of all. These fruits contain both a soft outer skin and a soft fleshy portion around its seeds. By contrast, dry fruits provide maximum protection to their seeds. Dry fruits differ so much from fruits such as peaches, you may not even think of these as fruits, but they are:

> ➤ Beans
>
> ➤ Walnuts and other nuts

> ➤ Peas and other pod fruits

> ➤ Grasses such as rice, wheat, and corn

> ➤ Milkweed

> ➤ Cockleburs

In addition to providing protection to the seeds, fruits help disperse seeds. They do this by being a food for birds and other animals. After eating the fruit, the seeds can often emerge undamaged from the digestive tract. Wildlife, in fact, plays an important role in seed dispersal from region to region. This aids in building genetic variation in the plant species.

An Angiosperm Life Cycle

The angiosperm life cycle begins with a flower's anthers and ovary. The anther produces a male reproductive cell that undergoes meiosis to become a haploid microspore. This microspore soon develops a hard outer coat and holds two male cells: one cell will be used in fertilization and the other, called a tube cell, is essential during pollination. You know this hardened pod containing two male cells as a pollen grain.

Meanwhile, meiosis takes place in the ovary, too. Meiosis in both the male and female parts of the flower produces a male gametophyte (the sperm cell) and a female gametophyte (the egg), respectively. The gametophytes are haploid cells. The ovary produces many ovules (eggs) but only a few are used for fertilization with the sperm. The rest become nonreproductive cells.

After pollen grains land on the stigma, the pollen's tube cell elongates and advances down the carpel toward the ovary. Fertilization can occur only after the tube reaches the ovules. The nuclei of both sperm cells target a single ovule. Fertilization occurs when a sperm nucleus merges with an egg nucleus. In a final critical step, the DNA from the sperm and egg merge. The newly formed embryo is the plant's zygote.

Remember that the pollen grain delivers two sperm cells in fertilization. Remember also that many of the ovules waste away and only one or a few remain as fully formed eggs, available to be fertilized. The extra sperm finds one of the extra eggs and they combine. As a result, they form a material called endosperm. Endosperm will enlarge as the zygote starts to develop into an embryo. The endosperm represents the food supply for the young growing plant.

Shortly after the fertilization steps, a seed coat develops. The seed coat offers the secure protection for the embryo as it grows and in the meantime preserves the seed. Many seeds can stay in a dormant form for years before coming to life when the conditions are right. The life cycle culminates when the seed germinates. The young seedling that emerges is a new sporophyte. When this sporophyte matures—you recognize it as a plant, bush, or tree—it will produce its own flowers and start the cycle again.

Biology Tidbits

Time to Bloom

Ever think about how a plant decides to unfurl its petals in the familiar action you know as blooming? A flower's new bloom is controlled by several hundred well-coordinated genes that are activated by specific levels of water, temperature, and sunlight. The signals from the environment can be very subtle and we surely do not understand all the information that a flower gathers from its surroundings before turning on the bloom genes. In addition to the main factors listed above, blooming may also be affected by winds, the plant's elevation, and other geographical features.

CHAPTER 28

Plant Sensory and Defense Systems

In This Chapter

➤ The things in nature that plants detect

➤ The basics of plant signal and response systems

➤ Plant hormones and what they do

➤ How plants defend themselves against predators and infection

In this chapter you will see how plants set up communication systems, sense their environment, and defend themselves against predators. Imagine you lived your entire life anchored to the ground by a strong root system. If you needed to flee from danger, you would have to uproot that anchor and disconnect yourself from the very soil that sustains you.

Of all the obvious differences between plants and animals, mobility may be the most impactful on the way these two groups of living things survive. Yet plants are not defenseless. They have ways to keep in touch with others of their species and fight off most, but not all, predators.

This chapter teaches you about two major aspects of plant life: (1) sensory systems for assessing the world around the plant, and (2) a plant's defenses against attack. I will begin with a section on the types of things in nature that plants sense.

Nature's Stimuli Sensed by Plants

A common feature in all living things is the ability to sense changes in environmental conditions and respond to those changes. Picture a chap reclining on a couch and watching a football game on television. He may not seem to be responding to much, beer in hand, eyelids droopy. But his body indeed responds to the announcer's call, a pizza commercial, pressure accumulating in his bladder, and subtle shifts in the light streaming into the room as the afternoon ages. Plants are pretty much the same; though they appear to be comatose, they actually operate a sophisticated system of sensory input, feedback, and response systems.

The main physical features of the environment sensed by plants are:

➤ Light intensity and direction

➤ Gravity

➤ Temperature

➤ Relative humidity

➤ Drought and flooding

➤ Salt levels in soil and water

➤ Mechanical forces caused by winds or by touch from people, animals, or equipment.

Biology Vocabulary

Apoptosis

In biology, apoptosis equals programmed cell death. Apoptosis occurs in microbes, animals, and plants, but it may be easiest for us to detect visually in the plant world. Certain genes get turned on in a plant when the environment changes. The genes initiate the breakdown of several substances that are important for actively growing cells, such as chlorophyll, DNA, proteins, and the lipids that keep membranes healthy.

You can see apoptosis in action when leaves change color in autumn or when a flower begins to die. The shutdown of certain vessels inside a tree, which will start the process in making wood, also relies on a complex set of apoptosis genes.

Apoptosis is a critical piece of plant survival. When leaves fall off *deciduous trees* in the fall, this action protects the tree from desiccation (drying out) during a long winter. Before the leaves fall off, the tree is busy recovering nutrients from these nutrients and storing them in the plant body for use later in the winter.

Plants monitor all of these physical traits of their environment almost constantly. By assessing these factors in specific combinations, plants determine when they should germinate their seeds, flower, release pollen, produce fruit, release seeds, go *dormant*, or start the process of cell death, called apoptosis.

A Plant's Signal Response System

A plant cell holds sensitive sensors on its surface for detecting changes in its world. Similar to the signal-response system in animals, a plant senses a physical feature in the environment. This is called a signal. The piece on the cell surface that senses the signal is called a receptor.

The plant's entire system for detecting the environment contains three main components:

> ➤ Reception—When the receptors detect a signal.

> ➤ Transduction—The process of sending information from the stimulated receptor to the cell's nucleus.

> ➤ Response—Enzymes receive the information and turn on certain genes. These genes in turn control the production of proteins that participate in building a response by the cell to the signal.

The response to signals in cells works exactly the same way as in animal cell protein production: It depends on transcription and translation (described in Chapter 13).

The most-studied signal response systems in plants relate to two key processes in plant physiology:

> ➤ Photoreception—A plant's process for detecting light of certain wavelengths as well as light intensity, direction, and duration.

Biology Tidbits

Photoreception in Plants

A plant without a photoreception system is a dead plant! Photoreception is essential for plant survival because photosynthesis depends on it.

Photoreception uses two main light receptors: blue-light photoreceptors and phytochromes. Light in the blue portion of the visible light spectrum activates the blue-light receptors. These receptors are involved with a plant's control of stomata for gas exchange and key steps in seed germination. Phytochromes also help regulate seed germination in addition to controlling steps in de-etiolation, plant bending to avoid shade, and responses to the daily light-dark cycle. Thus, phytochromes play an important part in giving plants a biological clock. In plants and other living things, the cumulative responses to a 24-hour cycle is called circadian rhythm.

➤ De-etiolation—A combination of actions taken by a plant when a stem shoot reaches sunlight and which starts the process called greening. Greening includes leaf growth, chlorophyll synthesis, root elongation, and a slowing of stem growth.

Biology Tidbits

Plant Hormones

Of all plant hormones, gibberellins play the biggest role in plant life. They were first discovered in the 1930s in the fungus *Gibberella*, a pathogen of rice plants. More than 100 different gibberellins have now been found in other microbes and flowering and non-flowering plants, each producing its own unique array.

Gibberellins stimulate division and elongation of leaf and stem cells, but have less effect on roots. Plant growers use gibberellins to make certain dwarf plant varieties reach normal size. Some fruit growers spray young plants with gibberellins to produce larger fruit. The large Thompson seedless grapes sold in groceries are the result of gibberellins.

Plant Hormones

Hormones are substances made by multicellular organisms in one part of their body and meant to work in a different part of the body. In animals, hormones circulate in the bloodstream until reaching the target tissue. In plants, hormones travel through the vessels and also pass across cell walls.

All hormones work in very low concentrations. In fact, a hormone's level in the body is usually not as important as its level in relation to other hormones. Hormone activity is usually the result of two or more hormones turning on a specific body function and another hormone that shows up to turn off the same function when the time is right.

Hormones operate by activating specific genes. These genes then control a particular behavior by the body. The main plant hormones are:

➤ Auxin—Stimulates root growth and cell differentiation in seedlings; regulates fruiting; controls xylem development; regulates falling of leaves (called leaf abscission); and responds to light and gravity.

➤ Cytokinins—Stimulates cell division and growth; stimulates seed germination; regulates root growth; and delays plant aging (called senescence).

➤ Gibberellins—Prompts seed and flower bud germination; stimulates stem, leaf, and flower growth and fruiting; and regulates root growth.

➤ Brassinosteroids—Inhibits root growth and leaf falling and promotes xylem development.

➤ Abscisic acid—Inhibits overall plant growth, closes stomata, and promotes seed dormancy, all in response to drought conditions.

➤ Ethylene—This gas initiates fruit ripening and affects root, stem, and flower development.

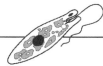

Inside a Cell

Circadian Rhythms

All eukaryotic cells possess some type of circadian rhythm. Circadian rhythms encompass all the processes inside a cell that makes the plant behave one way as night turns to day, during the daylight hours, and again when night returns. Circadian rhythms are built into a cell's genetic makeup so that certain processes automatically turn on when day comes and others turn on when night arrives. In this way, a cell does not have to say, "Oh gee, it's daytime again!" Its genes have already evolved to match its environment's seasons, climate, day length, and sunshine.

Fruit Growth and Ripening

An immature fruit is better at protecting the seeds inside it than mature or ripe fruit. Immature fruits are hard, tough, and tart. No bird, bear, or person would want to snack on them! But for a plant species to survive, it must disperse its seeds into the environment. The fruit eventually ripens and begins to attract animals. When animals eat the fruit, the seeds travel through the gut and come out the other end. Seed dispersal often occurs far away from where the fruit had been eaten. By this mutual relationship, plants and animals help each other survive.

Plants make fruit attractive to animals by developing bright colors and appealing scents. These attributes develop only at certain times of the year when seed dispersal will have its best chance of succeeding.

Ethylene is the main controller of this summertime process of fruit ripening. Ethylene activates the ripening process. More ripening prompts the release of even more ethylene. (This is a rare example in nature of positive feedback in which a substance makes the body produce even more of the same substance. Most feedback systems in biology involve negative feedback.) This gas also spreads to adjacent fruit so that they begin to ripen too. The saying, "One bad apple spoils the whole lot," is true, and we can blame ethylene for it.

A mature fruit carries the maximum amount of sugar for the seedling's use if the fruit happens to fall to the ground. Animals that eat the fruit take the sugar for themselves, but assuming a fruit manages to avoid being eaten, other physical changes help the seeds turn into seedlings:

> ➤ Water accumulation in the fruit's fleshy part.

> ➤ Softening of the protective skin or rind to help the seeds escape.

> ➤ Over-ripening or rotting caused by continued ethylene action, which further helps in seed dispersal.

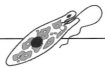

Inside a Cell

Seed Germination

A new generation of plants depends on reliable germination of a seed into a seedling that can struggle toward the sunlight and form a new plant. Seed germination has evolved to contain various controls that ensure a seed stays dormant until the conditions are perfect for germination. Plant hormones additionally regulate the steps of germination so that all goes as planned.

Germination proceeds in about four steps controlled mainly by gibberellins. Cytokinins and other hormones help regulate the steps. When a seed drinks in water, the embryo releases gibberellins. These hormones tell the embryo to make enzymes to digest the nutrient-rich seed endosperm. By using the nutrients, the embryo grows and soon elongates into a seedling. Once the seedling has emerged from the seed, ethylene takes over and helps control its upward growth through the topsoil toward sunlight.

Types of Plant Defenses

A tree deemed delectable by an African elephant on the savannah has few defenses, but plants manage to fight off lots of other much smaller predators. The main predators of plants are herbivores and pathogens.

Plants repel herbivores and pathogens by a combination of features that make the plant body unappealing, difficult to digest, or poisonous. Some plants even set up mutual relationships with certain animals (usually insects) that repel other animals.

Plant Defenses against Herbivores

A plant's best defense against being eaten is to become as unpalatable as possible. You have seen why immature fruit helps protect the seeds because of the fruit's disagreeable features. Plants use other defensive mechanisms that are equally effective:

➤ Thorns and spikes keep animals from biting into the plant body or its leaves.

➤ Distasteful substances repel animals and prevent them from eating more of the plant.

➤ Poisons repel or kill insect predators.

➤ Some plants attract certain insects that combat other predatory insects.

➤ Some plants under insect attack release into the air substances that signal neighboring plants to turn on their own defenses.

Plant defenses must achieve a balance between predation and the benefits of plant-animal relationships. Plants still need animals for seed dispersal and pollination so they cannot make their defenses so strong that they harm the plant's chances for survival.

Plant Defenses against Pathogens

Bacteria, viruses, and fungi attack plants through the roots, the leaves' stomata, and injuries to any part of the plant body. Plants defend against some attack by building strong cell walls, but they cannot stop a pathogen that has found its way into the body through an opening.

Biology Tidbits

Salicylic Acid

In plants, salicylic acid activates a type of immunity against infection called systemic acquired resistance (SAR). SAR protects plants from future repeat infections by the same pathogen that is causing an ongoing infection. Centuries ago, chemists learned that they could modify the salicylic acid molecule in their laboratories to make aspirin. Swallowing an tablet of aspirin could lessen certain pains such as headache or toothache. Some societies had already discovered the pain-relieving benefits of plant extracts. For example, by chewing the bark from a willow tree, minor pains would temporarily disappear. We now know the bark contains high levels of salicylic acid.

Plants fight infection in ways similar to an animal's immune defense against infection:

➤ Natural resistance to certain pathogens developed during a plant's evolution.

➤ Substances released when the cell wall is infiltrated or damaged stimulate production of *antimicrobial* compounds. These antimicrobial compounds are called phytoalexins.

➤ Immune hormones protect the plant against repeat infections by the same pathogen. Salicylic acid is a plant substance that helps plants develop this long-term immunity.

PART EIGHT

Environment
and Ecology

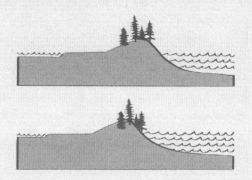

CHAPTER 29

 **Environments**

In This Chapter

➤ The concept of environment and Earth's main environments

➤ Descriptions of the main soil and water environments

➤ How ecosystems, environment, and living things relate

In this chapter and the next the world of biology branches out to you show how biological things interact with the nonliving world. We use the word "environment" repeatedly these days. Perhaps because of this possible overuse, environment has lost its meaning. This chapter discusses the features of the environments where biological things live. It shows how certain physical factors of the environment make life possible.

To understand biology in the big picture sense, it helps to understand the meanings of two closely related areas of science: environmental science and ecology. These two fields have become almost interchangeable, but each has a different focus when you get into the details.

Environmental science draws on features of the physical Earth to figure out how life exists and how the Earth works. Ecology is the study of how living organisms interact with each other and with their nonliving environment. An environmental scientist might study the composition of soil or water. During the study, the scientist would also examine the organisms thriving in a clump of soil or a teaspoon of pond water.

Ecologists go straight for the ways in which soil microbes interact with each other and the ways microbial populations interact with soil invertebrates. The ecologist spends equal time studying the invertebrates' relationship with birds, mammals, reptiles, etc. Finally, ecology concerns the ways in which all of these interactions involve nonliving matter: the rocks, soil, water, and air that compose the environment.

Biology Vocabulary

Environment

We use the word "environment" so often these days, we may have lost its meaning. In biology and the earth sciences, environment consists of all the external living and nonliving conditions that affect an organism in its lifetime. By this definition, environment includes your habitat's climate, weather, geology, geography, plant and animal life, and the soil, water, and air that comes in contact with you.

The Earth's Environments

The nonliving Earth can be divided into three major environments: soil, water, and air. Soil and water environments provide a habitat for far more organisms than the air. Even though the air contains many microbes, very few of these microbes live their entire life cycle in the air. The air serves mainly as a transport medium for bacteria, viruses, fungi, insects, and pollen. By contrast, numerous organisms in soil and water grow, metabolize nutrients, reproduce, and form communities in the soil and water habitats. Most important, the organisms of soil and water occupy a niche wherein they carry out a job that benefits other organisms in the same environment.

Biology Vocabulary

Habitat and Niche

Every living thing occupies a favored habitat and a particular niche in an ecosystem. Habitat is the place where an organism normally lives, such as a marsh, boreal forest, or desert. A niche is the role played by an organism in its ecosystem. In a pond ecosystem, for example, algae occupy a niche related to photosynthesis. Specifically, algae and other photosynthetic organisms in the pond make energy available to every other organism in the pond.

Within soil and water, many additional and more specific environments exist. Each of these environments support distinct populations of organisms from microbes to higher multicellular organisms:

➤ Surface soil or topsoil—The uppermost several inches of soil.

➤ Deep soils—Starts below the topsoil and extends to many feet deep.

➤ Deep sediments—The soil under the deep soils extending to miles from the soil surface and defined by complete darkness, low oxygen levels, and limited nutrient supply.

➤ Surface freshwaters—The uppermost layer of rivers, lakes, and ponds that receive sunlight.

➤ Deep freshwaters—Freshwater in deep regions where little or no light penetrates and extending to the bottom. Characterized by cold temperatures and limited nutrient supply.

➤ Surface marine waters—The uppermost layer of the ocean, bays, and estuaries that receive sunlight.

➤ Deep marine waters—Marine waters in deep regions where little or no light penetrates and extending to the bottom. Characterized by cold temperatures and limited nutrient supply.

Each of these environments can be further broken down into more specific environments, such as clay soils, sandy soils, marsh water, and a lake bottom, to name a few. When an environment has very specific conditions or lies in a limited area, scientists call the environment a microenvironment. By breaking up the big Earth environment into small pieces, scientists can focus on the specific conditions that support life in each of these places.

Biology Vocabulary

Microenvironment

A microenvironment is an environment that is narrowly defined in terms of physical features (temperature, light, moisture, etc.) or chemical conditions (acidity, nutrients, oxygen levels, etc.). Examples of microenvironments are: the inside of the rumen, the rhizosphere, and a hot sulfur spring.

Soil Environments

Soil is a large ecosystem composed of several more narrowly defined ecosystems. Soil composition shifts from place to place such as the soils of old growth forests, woodlands, grasslands, marshes, and deserts. Consider only the soil under a pine forest. It contains several sub-environments:

> ➤ Leaf litter that covers the topsoil.

> ➤ Topsoil, a porous mixture of partially decomposed leaves and other organic matter, inorganic particles, and the microbes, insects, and animal and plant life living there. The topsoil holds the majority of plant roots but tree roots extend deeper. The majority of animal and plant life occurs in this layer.

> ➤ Subsoil, a more tightly packed soil where tree roots extend and where water percolates down from the top layers.

> ➤ Parent material, the support layer or bedrock of soil. Tree roots stop here and most animal and plant life disappears. Microbes exist on limited nutrients and low oxygen levels.

Soil scientists delve even deeper into each of the environments listed above. Thus, some scientists devote careers looking only at the conditions found in sediments of the sea bottom, deep soils of the Earth's crust and mantle, the rhizosphere, and even the tiny pockets of space between soil particles.

Water Environments

Water environments contain many layers similar to how soil can be layered. Though water has more mixing between layers than soil, its sub-environments remain remarkably consistent over long time periods. For ease, I present first the layers or zones found in lakes and many deep rivers, then follow with a list of zones of the ocean.

> ➤ Littoral zone—Surface water near shore that receives large amounts of sunlight. Although this zone can be very shallow, the roots of terrestrial plants stop growing here.

> ➤ Limnetic zone—The open, sunlit surface water away from shore. Its depth is determined by how far sunlight penetrates to keep alive the photosynthetic organisms of this zone.

> ➤ Profundal zone—The deep open water too dark to support photosynthesis. The water here is always cool and contains fish adapted to dark and cold water.

> ➤ Benthic zone—The sediments of the lake bottom and the slopes toward shore. The invertebrates and microbes living here depend on nutrients from decayed material that falls from above or comes into the lake in runoff from the land.

Rivers have the above zones, but the overall surface environment of a river can be divided into three additional zones. Each river zone supports distinct plant and animal life:

➤ Source zone—At high elevations where the flow begins from springs, lakes, glaciers, rain, or snowmelt.

➤ Transition zone—The region where tributaries feed the river and enlarge it. Nutrients flow into the water that feed both life in the river of this zone and downstream.

➤ Floodplain zone—Here, the river swells and flow slows. The river may create lakes, marshes, or a delta. Sediments wash toward the ocean, and at some point, freshwater and salt water mixes.

The ocean possesses similar stratification. The marine layers relate mainly to amount of light they receive and the water's temperature, but they also vary by nutrient and oxygen levels and pressure. As shown below, the ocean zones occur in both horizontal and vertical directions:

➤ Coastal zone—Water from the beach to the edge of the *continental shelf*. This zone contains abundant animal and photosynthetic plant life, about 90 percent of all the world's marine species.

➤ Open sea zone—All the water beyond the edge of the continental shelf, from the surface to the sea bottom.

➤ Euphotic zone—All water, coastal or open sea, from the surface to a depth of about 200 meters (656 feet) where photosynthesis remains active.

➤ Bathyal or mesopelagic zone—Wonderfully nicknamed the "twilight zone," this layer ranges from 200 meters to about 1,500 meters (4,921 feet) deep and receives low levels of filtered light. This zone has cool water, reduced oxygen levels, and increasing pressure. Some specially adapted microbes and plants can catch enough light to run photosynthesis. Small organisms living here depend on the food that drifts down from upper layers. For larger species, it's a fish-eat-fish world where every living thing is hungry.

➤ Abyssal zone—This region below 1,500 meters to the sea bottom is one of eternal darkness and frigid waters. Nutrients are scarce here and life depends on a thin stream of food particles from above or the occasional carcass of a large fish or marine mammal. Organisms in this zone are adapted to the extreme hydrostatic pressures the ocean exerts.

Certain regions of the ocean floor contain hydrothermal vents. These underwater geysers spew super-hot water from the Earth's mantle. This water boils away as soon as it hits the ocean water, but it pours forth a mixture of minerals and acid that creates a living community like none other on Earth.

Biology Tidbits

Hydrothermal Vents

Hydrothermal vents occur in the Pacific and Atlantic Oceans along a marine mountain range called the Mid-Ocean Ridge. A specialized type of hydrothermal vent called black smokers release outpourings of iron and sulfur compounds, and these vents are characterized by black emissions rather than the gray material from other hydrothermal vents. A limited community of organisms can withstand the enormous pressures (of greater than 300 atmospheres), temperatures reaching 700° F, and the caustic chemicals at the vent's rim.

Archaea and specially adapted invertebrates live here. One invertebrate is a type of tubeworm that grows to eight feet long and has no eyes, mouth, or stomach! This organism survives by depending on the microbes that live inside its tissues. The microbes live on the minerals from the vent and convert them into cell constituents. The tubeworms, which cluster around a vent's forbidding mouth, get their nutrition from the microbial cell materials.

Ecosystems

An ecosystem is a defined place where plants and animals interact with each other and with the nonliving things in their environment. Ecosystems can be huge, such as the ocean. Many ecosystems take up a smaller space, such as a single lake.

Environmental scientists also think of ecosystems in both an open-ended sense and in a narrow sense. For example, a scientist might refer to all soil on Earth as the soil ecosystem. Another scientist might be more specific and consider only the soil of deciduous forests of North America to be an ecosystem. Yet another scientist could define a certain ecosystem to be the soil of a particular grassland in Kansas. All are correct. The point of defining an ecosystem is to identify the exact animal and plant species using the place, interacting with each other, and interacting with the physical features found there.

Ecosystems often work in very specific ways. As mentioned, they contain a fairly defined group of species that can always be found there. Some ecosystems contain only microbes, such as very deep sediments, but most Earth ecosystems consist of a wide variety of microbial, plant, and animal life.

One of an ecosystem's defining features relates to its nutrient flow. In ecosystems of surface soil and water, photosynthetic microbes and plants supply energy to other living things by capturing solar energy. Certain animals then eat the plants, and other animals eat the first animals. This flow of nutrients and energy from plants to animals to another series of animals is a food chain. When several food chains in an ecosystem crisscross, they form a food web.

Biology Vocabulary

Food Chains and Food Webs

A food chain is a series of organisms in which one eats the preceding one. For example, a food chain might start with sea grass, which is grazed by marine invertebrates, which are eaten by fish, consumed by seals, and devoured by a shark. The shark is said to occupy the top of the food chain (because not many things want to take on a shark!)

A food web consists of a complex network of interconnected food chains. A small pond might hold a very complex food web starting with photosynthetic algae that feed protozoa, then invertebrates, then small fish, and finally larger fish. But the pond contains a variety of fish, some of which eat each other as well as the invertebrates. Protozoa live inside several species and help with digestion. As additional food chains develop in the ecosystem, you begin to see how nature is made of species all interdependent on each other.

CHAPTER 30

Ecology

In This Chapter

➤ How humans fit into Earth ecology

➤ How nutrients cycle through Earth's ecosystems

➤ Biodiversity and why it is important

In this chapter you will see how you fit into the Earth's overall biology. This study of how living things relate to each other and to their environment is called ecology.

Never before has ecology been in the forefront of our thoughts as much as it is now. Our time on Earth has been defined by a heavily populated planet, disappearing resources, mass extinctions of plants and animals, and global climate change. We know humans impact the environment and have dramatically influenced ecosystems and Earth ecology. We may argue about the degree to which humanity affects these things, but few people could doubt that pavement, bridges, runways, cities, and mass changes in forests, grasslands, and coasts are affecting nature.

Today's ecologists study not only undisturbed ecosystems (if they can find any left!) but also the effects of human activities on existing ecosystems.

How Humans Fit into Earth Ecology

Do people participate in ecosystems? It seems as if we have effectively separated ourselves from nature. We live in houses sealed off from nature, commute in vehicles on strictly defined thoroughfares and flight paths, and go to work and school in yet another building where windows are locked and light is often artificial. An ecosystem in a babbling brook outside your window, in a treetop, or at the beach seems to have little in common with human life.

Human society of industrialized countries does not interact with ecosystems as much as ancient hunters-gatherers or people now living subsistence lifestyles. In parts of the world, many people hunt for game almost daily, pull fish from a nearby shore, and clear a section of forest to plant seeds. By doing these things, they take an active role in ecosystems.

Industrialized society influences ecosystems too, but indirectly. This society changes the makeup of ecosystems by putting pollution and other wastes into the air, water, and soil. These materials change plant and animal nutrition. Enough waste buildup can crowd a species out of an ecosystem that it had occupied throughout its evolution. When we think of how humans affect ecosystems, their influence is, sadly, often one of pollution, destruction, *imbalance*, or habitat fragmentation.

Humans have their greatest influence on the environment and ecology by participating in biological communities. A community is all the populations of different species—plants, animals, and microbes—that live and interact in a certain area. For example, a community exists on the African savannah. Another different community occupies boreal forests. Still other communities flourish on a barrier island off North Carolina's coast or a coral reef in the South Pacific. How people live their lives in these places impacts thousands of other species directly or indirectly.

Biology Vocabulary

Species, Populations, and Communities

Thinking about biology in terms of ecology and ecosystems can be complicated. But at their most basic, these things are derived from species, populations, and communities.

A species is a group of organisms that are all similar in appearance, behavior, chemistry, and their genetic makeup. A population is a large number of individuals all of the same species and which are interacting. A community is a group of several different populations living in the same area and interacting.

Human Activities and Ecology

Human activities affect ecology. No one can argue with that because every activity of every species on Earth affects ecology in some way. Unfortunately, it seems humans do more things that hurt ecosystems than help them. Some of the ways humans potentially harm ecosystems are:

➤ Waste and pollution—Can affect the normal nutrition of species, poison some species, and force other species to leave a habitat or go extinct.

➤ Population spread and urbanization—Remove habitat from species that need it, and also lead to overfarming, overgrazing, and overfishing of specific regions.

➤ Noise and light—Affect normal behavior of many species as well as migration and breeding patterns.

➤ Climate change—Wastes from human activities entering the atmosphere cause it to trap the Sun's energy at higher-than-normal levels. An overall warming of the Earth's surface shifts normal weather patterns, plant and animal reproduction, feeding behavior, and water supply.

Before you think I'm blaming humans for the destruction of biology, I emphasize here some of the ways people have tried to help the Earth's ecology. Sometimes we make mistakes or an approach to saving the planet turns out to cause problems years later. Fixing a planet is no easy job, but the following actions show that people have tried, and continue to try, to slow damage already done to ecosystems:

➤ The three R's: Reduce, Reuse, and Recycle—Waste reduction and the recycling of resources help lessens the amount of overall waste we put into the environment.

➤ Renewable energy—Increasing dependence on the energy carried by sunlight, wind, water (tides), and geothermal sources and decreasing dependence on energy from fossil fuels (coal, oil, and natural gas), which contribute to *global warming*.

➤ Conservation—This refers to plant and animal conservation as well as land conservation and habitat protections for endangered and threatened species.

➤ Environmental legislation—Laws intended to protect species and their habitats, thus slowing or avoiding their extinction and helping maintain ecosystem vitality.

How Species Interact

Why does the protection of species living halfway around the world and unknown to most of us matter? Think again of the community concept. All species in a particular region interact. They do this by (1) making nutrients available to each other; (2) shaping the land by digging, burrowing, cutting, or damming; and (3) keeping populations under control so that all individuals have a fair shot at food, water, and shelter. (This last point is discussed further in the next section on prey and predators.)

You could probably think of several ways in which animals and some plants help the environment. Here are some examples:

➤ Coyotes keep rodent populations in check.

➤ Rabbits convert plant energy to a food source for the *carnivores* that eat them.

➤ Phytoplankton converts the Sun's energy to a form that supports food chains.

Biology Tidbits

Energy Pyramids

Nutrients cycle, but energy only flows. It furthermore flows mainly in one direction. To begin energy flow, Earth's plants convert a massive amount of solar energy to plant energy. Large populations of *herbivores* take in this plant energy and store it in their animal tissue. Animals that eat the herbivores get energy, and if another species should prey on them, the predator would get some of the energy too.

This flow of energy is called a pyramid for two reasons. First, the amount of photosynthetic plants exceeds the total mass of animals that graze on them. Herbivores similarly outnumber their predators. The predator atop the pyramid produces much smaller populations than its prey. Think of this: blades of grass far outnumber the caterpillars that chew on grass; caterpillars outnumber the songbirds that eat them; and songbirds outnumber the hawks that prey on songbirds.

The second reason relates to a similar flow of energy. Almost all the solar energy that the Earth can absorb resides in photosynthetic organisms. But eating them causes a slight loss of the energy. Animals and their predators can never recover all the energy from their meal. So some energy is always lost (as heat) as we move up the pyramid from plants to herbivores to prey and predators.

Nutrients Cycle through Earth's Ecosystems

Microbes in soil and water make a tremendous contribution to the environment. They do this silently and unseen by you, but their activity continues every minute of every day. This contribution to ecology is called nutrient cycling.

Bacteria and fungi employ their enzymes to break down dead plant and animal tissue to their component molecules. These molecules seep into the earth where other microbes reassemble them into cell materials. Large microbes eat small ones; tiny invertebrates eat the microbes; insects eat the invertebrates; and up the food chain the nutrients go!

As certain nutrients pass through living and nonliving matter, they enter the atmosphere. Different organisms recapture the nutrients by respiring (oxygen), photosynthesizing (carbon dioxide), or performing nitrogen fixation.

Each nutrient cycle consists of several steps carried out by different microbes. These nutrient cycles also have stages in which the nutrient goes through different forms as it passes through the soil, air, a microbe, a plant, or an animal:

> ➤ Nitrogen cycle
>
> ➤ Carbon cycle
>
> ➤ Sulfur cycle
>
> ➤ Phosphorus cycle
>
> ➤ Iron cycle
>
> ➤ Oxygen cycle

In these cycles, a microbe puts the nutrient in a chemical form that can be used by another microbe or a higher organism. When animals eat plants or prey on another animal, they move the nutrient along in its cycle.

A major reason for ecosystems is the recycling of nutrients so that a maximum number of species can benefit from it. This recycling allows biological things to sustain their populations on Earth for generations, rather than using up a resource so that it is forever gone.

Prey and Predators

In predation, animals of one species feed on members of another species. (Insert your favorite Wall Street joke here.) The benefit of prey to the predator seems obvious: by eating prey, a predator stays alive. Predators also benefit their prey by removing weak or sick members. This helps the prey strengthen its population's genetic makeup. Predation also keeps prey populations from growing too large and using up all their food and space.

It's hard not to root for a seal pup when being chased by a shark, but predators such as sharks contribute to ecosystem health in ways we know and probably in many ways that have not yet been discovered.

Biodiversity

Biodiversity is the variety of life that is needed to sustain a particular ecosystem or community. Biologists study different types of biodiversity. You may be most familiar with species biodiversity, which is the variety of different species in a location on Earth. But biodiversity can mean these additional types of variety:

> ➤ Genetic diversity—Variety in the genetic makeup among members of the same species.

> ➤ Ecological diversity—Variety of ecosystems in a particular region.

Biodiversity declines when species go extinct. The Earth's species need biodiversity to support the multitude of food webs that exist in ecosystems. A decrease in biodiversity lessens the chances for any particular species' ability to get food, avoid predators, build shelter, and prepare for a new generation of offspring.

Biodiversity benefits humans in hundreds of ways we know and possibly thousands of ways we have not discovered. People use substances made by plants and animals for food, drugs, antibiotics, clothing, and industrial products. When species go extinct, they disappear forever with any potential benefits they held for humanity. Of course, biodiversity does not necessarily exist just for human benefit. With biodiversity, the world is a rich, interesting place to live. Without biodiversity, species struggle. With biodiversity, humanity and the species that share nature with us all profit.

Biology Vocabulary

Extinct and Threatened Species

An extinct species is one that has completely disappeared from Earth. Extinction happens when a species cannot adapt to changes in the environment or reproduce fast enough to maintain its population size.

A threatened species is one that may still be abundant but is in danger of heading toward extinction because its numbers are declining.

Biodiversity and Ecosystems

Ecosystems work best when they contain many different species with numerous interactions. This complexity protects the ecosystem in case of natural disaster. Should some species be wiped out in the disaster, other species would take over some of the same roles in nutrient cycling and in food webs.

An ecosystem built on few species is much more vulnerable to destruction. Perhaps each species carries out a certain role. Lose that species to a flood or a drought, and the other species that depended on it may not be able to survive.

Diversity strengthens and insures the survival of all biological systems.

Climate and Biodiversity

Climate refers to the weather patterns and atmospheric conditions that characterize a region over a long time period. The Earth contains local climates in addition to a global climate.

The species inhabiting the Earth today evolved due to adaptations to local climates and the general ability to survive in the global climate. Thus, a region's ecosystems have developed based on certain temperature ranges, rainfall amounts, and food supply.

The global warming associated with today's progressive climate change shifts a region's conditions. Every species on Earth has an *optimal range* of conditions in which it thrives. Outside this range, it might not survive. Even conditions toward the outside of this optimal range cause a species to struggle. The effects of global warming may seem subtle to us, but they are having fatal and near-fatal consequences in the populations of many other species.

Deforestation and Desertification

Two aspects of climate change relate to deforestation and desertification. Deforestation is the removal of trees from a forested area without adequate replanting to maintain the trees' numbers. Desertification is the conversion of grassland and other open spaces to desert-like conditions that cannot sustain agriculture.

People directly cause deforestation by cutting down trees and failing to replace them with young trees. The lumber, paper, road-building, and real estate industries have been blamed the most for deforestation. Poverty also contributes to deforestation when villages burn forested land because they need the land for growing crops. Fires and other natural disasters do not cause major losses to forests compared with human activities.

Biology Vocabulary

Habitat Fragmentation

Habitat fragmentation is the breaking up of a species natural habitat, range, or migration routes into smaller pieces. Human activities are often the cause of habitat fragmentation. The main causes are: new housing developments, golf course and airport building, fences, and roadways.

Fragmented habitats harm species for various reasons: (1) decreased space for hunting and shelter; (2) overcrowding of populations in smaller areas, which leads to increased disease and fatalities due to territorial fights; (3) inadequate water; (4) shortage in the number of places or materials for building nests or dens; and (5) weakening of genetic diversity because breeding populations are cut off from each other.

Desertification is the result of poor land management and global warming. Land usually becomes unsuitable for producing an adequate crop yield because of four factors: (1) overgrazing, (2) soil erosion, (3) prolonged drought, and (4) climate change.

CHAPTER 31

Biodiversity

> ## In This Chapter
>
> ➤ The world's biodiversity hotspots
>
> ➤ Ways that scientists try to save biodiversity
>
> ➤ Today's main threats to biodiversity
>
> ➤ The value of biodiversity to humans and nature.

In this chapter we continue the discussion of a most important aspect of biology, called biodiversity. All aspects of biology come together to make up the Earth's biodiversity. This is such a big subject that people often have a hard time figuring out how we fit into the larger world of diversity around us. Living inside houses and working inside office buildings might not help. Humans have become disconnected with nature and, because of this, we fail to see the ways we relate to other living things and the ways in which we differ.

A century or more ago, few scientists strayed outside their narrow field of interest. Even graduate students studying as recently as a few decades ago did not consider physics if their subject was biology. Geologists had no interest in plant reproduction. Now, even nonscientists realize that the sciences are interconnected; a typhoon in the South China Sea might affect fish populations in the North Atlantic.

People have been reluctant to admit the impact of many human technologies on the environment, but a generation of young students knows we can never again separate the nonliving and living worlds.

Biodiversity is the variety and number of biota in any particular region on Earth. It is the product of billions of years of Earth formation and the advance of independently living things. You cannot understand biodiversity without knowing a little something of biology, and at this time and forever forward, you can't consider biology without also discussing biodiversity.

This chapter presents the subject of biology in the context of biodiversity.

The World's Biodiversity Hotspots

Certain regions of the world hold an astounding variety of microbe, animal, and plant species. In these areas, both the variety of species and the number of species are very high but, also, the threats to these species are particularly high. Ecologists call these places biodiversity hotspots.

Biologists began to designate the world's hotspots in the 1980s. About that time they realized that species were disappearing at an alarming rate and that science might not be able to save every species that was going extinct. To forestall the premature extinction of as many species as possible, scientists decided to focus on the hotspots. By saving a hotspot, humanity could save a larger number and variety of species than if people tried to rescue every threatened region all at once.

Do you notice how I am emphasizing the places where the species live? This is a central theme of biodiversity: you cannot save a species unless you preserve its habitat. Unlike humans that seem to adapt to skyscraper apartment buildings as well as they adapt to mountain cabins, many of the species now in the most perilous situation are in danger because their habitat is almost gone. Habitat loss is inextricably linked to biodiversity loss.

Biology Vocabulary

Premature Extinction

Every species on Earth has either already gone extinct or will go extinct. That's the way evolution works, even for humans. Today's concern about biodiversity loss relates to premature extinction, which is the disappearance of a species at a rate faster than its natural rate of extinction.

About 25 hotspots exist worldwide. They occur on every continent except the polar regions. To qualify as a hotspot, conservation groups want a region to meet two requirements. First, the region should currently contain at least 1,500 vascular plants and, second, the region should already have lost 70 percent of its original area. Vascular plant numbers work in qualifying hotspots because many other species depend on plants for their survival.

Endemism

Scientists try to prioritize the hotspots that are in the most need of recovery. One measure of prioritizing hotspots is by determining a quality called endemism. Endemism is the degree to which a species is found in only one place on Earth. For example, cockroaches have a low degree of endemism. By contrast, snow leopards, which are limited to mountain ranges of Central and South Asia, have a higher degree of endemism.

Ways of Protecting Biodiversity

Ecologists take two main approaches to protecting biodiversity in hotspots. The first method focuses on particular species to protect. Biologists design recovery and protection plans for that species, including the plant's or animal's habitat. The second method considers whole ecosystems. By this approach, biologists design a plan to protect the land, water, and air that support a community of species. Protecting a marsh, for instance, would be an ecosystem approach to biodiversity protection. By preserving a marsh, people protect the soil and water that compose it. The microbes, animals, and plants that depend on those habitats can then continue interacting in their marsh ecosystem.

Threats to Biodiversity

The best ways to save biodiversity is to protect habitat. In turn, the best way to protect habitat is to eliminate as many of the immediate threats to the habitat as possible. After eliminating immediate threats, people can tackle more difficult-to-solve threats that underlie many of our current problems with the environment.

Immediate threats to biodiversity are:

➤ Habitat loss.

➤ Habitat fragmentation—Splitting a single large habitat into smaller disconnected pieces.

➤ Pollution of ecosystems.

➤ Influx of nonnative species that push out native species.

➤ Removal of keystone species.

After solving the above problems, people should consider the underlying problems in society that affect Earth biology:

➤ Poverty—This may be the single biggest threat to habitats and the hardest problem to solve. Impoverished societies turn habitat into land for crops or grazing, or remove resources to earn money. Illegal poaching, overfishing, and the sale of exotic species are by-products of poverty in parts of the world.

➤ Fuels—Burning of fossil fuels contributes to global warming, which directly threatens the health of plant and animal species. Wood also serves as the main fuel for heating and cooking in a large portion of the population. Cutting down forests for this fuel has devastated habitats and also added to global warming by removing the trees that pull carbon dioxide from the air.

➤ Climate change—All of the consequences of climate change have not yet been determined. We know that global warming increases ocean temperature, air temperature, weather patterns, and the melting of polar ice. All of these changes are occurring faster than many species can adapt. The failure to adapt to changes in the environment leads to premature extinction.

➤ Overpopulation—This problem affects biodiversity directly when habitat is lost to expanding urban growth and pollution harms ecosystems. Overpopulation also plays a hand in climate change and the levels of poverty ingrained in certain regions of the world.

Biology Vocabulary

Keystone Species

A keystone species is one that affects many other species in an ecosystem. The term "keystone" refers to the manner in which other species' survival relies on the presence of this single, critical member of the ecosystem.

The gray wolf of North America plays the role of keystone by culling weak animals from herds of elk, deer, and caribou. Population control of these herbivores gives young vegetation a chance to survive rather than succumb to overgrazing. Meanwhile, carcasses from wolf kills feed bald eagles and other scavengers. This in turn protects small mammals, birds, and fish.

The Value of Biodiversity

Can you put a value on biodiversity? Most people value biodiversity based on what it can do for us. The world's diverse species hold a treasure trove of medicines and products for human use.

Hundreds of foods come from the diversity of plant and animal life the world over. Antibiotics and anti-cancer drugs come from nature. Any synthetic forms of these drugs we now use are made by chemists using a blueprint already designed in nature. Many industrial chemicals, clothing materials, and building materials also come from diverse plants and animals.

Biodiversity also gives less tangible benefits to people. We enjoy biodiversity when we appreciate the solitude of a forest, peer into a tide pool, or even visit a zoo.

Humans are a single species among about 20 million total species of life on Earth. Biodiversity undoubtedly benefits those other species in ways we know and ways we cannot yet fathom. Biodiversity supports food webs that in turn support ecosystems. Removing even one species from an ecosystem can destroy the complex relationships that make up the ecosystem.

Indicator Species

Biologists try their best to monitor the health of ecosystems by watching indicator species. An indicator species is one whose well-being gives an indication of the overall health of its ecosystem. For example, the northern spotted owl serves as an indicator species of old-growth forests of the Pacific Northwest. Because this owl has very specific requirements for nesting and food, any change in its population indicates a possible change in the forest's condition.

Indicator species cannot tell the whole story on ecosystem health and biodiversity. Because it is impossible with today's technology to monitor every species in an ecosystem, biologists learn what they can from indicators. They then make their best guesses about what might be going on in the rest of the ecosystem.

We may never know the full extent of biodiversity on Earth. Hundreds of species will disappear without our knowledge. Those species take with them their genes and their contributions to ecosystem health. Scientists do not know all the reasons why we must protect biodiversity; they only know that once biodiversity disappears, replacing it is all but impossible.

CHAPTER 32

Today's Biology

In This Chapter

➤ An overview of specialty areas in biology

➤ Introduction to genomics and proteomics

➤ How conservation biology is using genomics to slow biodiversity loss

---

In this chapter you may consider the current status of biology in today's world. Biologists help solve problems in medicine and health, the environment, food production, and even new approaches in manufacturing. Biology continues to start with a single cell, but it extends so far beyond the cell in today's technology that the scope of this field becomes difficult to comprehend.

Perhaps the hallmark of what scientists call the "new biology" is its breadth. Biologists now study the tiniest inner workings of a virus. Across the street, another biologist may be contemplating the Earth's biodiversity or life on other planets.

This chapter reviews for you the many avenues in biology today. It then describes three major areas of interest: molecular biology, genomics, and proteomics. They may sound like fancy names, but these specialties have become part of almost all aspects of life from agriculture to vaccines. Each of these fields also contributes to a growing area of biology called conservation biology. Conservation biology applies the newest technologies of biology to solve problems in biodiversity loss. I will discuss conservation biology in this chapter.

Different Types of Biology

Biology has diverged into dozens of specialties. These specialties connect with each other as well as to different sciences such as chemistry. Perhaps we should never again think of sciences as unrelated. They all connect in their goal to explain how the universe works.

The main specialties in biology are listed below. Most of these can be applied to animals and plants. Each specialty has detailed sub-specialties. For example, microbiology includes bacteriology, mycology, virology, and protozoology. You will never run out of books to read on biology.

➤ Cell biology—The science of eukaryotic cell physiology.

➤ Molecular biology—The science of the workings of a cell's genes.

➤ Genetic engineering—The science of putting two different types of DNA together.

➤ Biochemistry—The science of chemical reactions of living cells.

➤ Microbiology—The science of microscopic organisms.

➤ Parasitology—The science of the parasites of higher animals.

➤ Genetics—The science of heredity.

➤ Genomics—The science of all the genetic matter of an organism, called its genome.

➤ Proteomics—The science of an organism's genetic matter and the proteins associated with it.

➤ Immunology—The science of how the body fights infection.

➤ Physiology—The science of an organism's overall metabolism.

➤ Anatomy—The science of how body parts work and their structure.

➤ Systems biology—The science of the systems that make up organelles, cells, or organisms and their interactions.

➤ Evolution—The science of how life developed on Earth.

➤ Developmental biology—The science of how organisms develop from an embryo.

Look at the nature around you right this minute. Pick any living thing. Somewhere in the world, a biologist is studying that living thing in detail you could hardly imagine.

Molecular Biology

Molecular biology targets only the biology inside a cell. In this field, scientists exclude the workings of tissues, organs, and organisms. Molecular biologists prefer to study a cell's genes, DNA and RNA, proteins and other constituents, membranes, and chemical reactions.

Molecular biologists may be credited with expanding our understanding of how DNA replicates and how transcription and translation work. These discoveries have been instrumental in finding cures for many inherited diseases. Watson and Crick were molecular biologists before anyone had ever used that title. Today's molecular biologists continue the study of DNA structure and the structure of other macromolecules of the cell.

Molecular biology includes the specialization called bioengineering. Bioengineers take DNA out of two unrelated cell types and combine them to make a new type of DNA molecule. Many people worry about the ramifications of making a DNA molecule that has never before appeared in nature.

Biology Vocabulary

Bioengineering

Several years ago, lab researchers shortened the term "biological engineering" to the catchier "bioengineering." In this science, a biologist uses an enzyme to cleave an organism's DNA. The scientist then mixes in a gene from an unrelated organism and employs another enzyme to sew the new gene into the DNA. The result is a new, hybrid DNA. When put into an organism, such as a microbe, plant, or animal, the hybrid DNA can change the recipient's normal physiology.

Genomics

Genomics is the study of all aspects of a cell's genetic material. This includes how DNA replicates, how genes are organized on chromosomes, and how all the information in genes winds up in proteins that run an organism.

Perhaps the biggest contribution of genomics to our present knowledge of biology relates to gene sequencing. In gene sequencing, a scientist determines the exact order of each nucleotide of an organism's DNA. For example, the Human Genome Project of the 1990s employed many scientists who collaborated to identify all the genes of the human genome.

What does "identifying a gene" mean? A gene has been identified when a scientist completes three steps:

> ➤ Find a gene on a particular chromosome.

> ➤ Determine the sequence of exact nucleotides that form the gene.

> ➤ Determine what the gene does.

The main way to determine what a gene does is to determine the protein associated with this gene. The best way to identify a protein is to test it in a lab and figure out what activities it controls in a cell. But this can be a very difficult undertaking. The science of proteomics developed mainly because of biology's need to learn more about the gene-protein connection.

Biology Vocabulary

Genome Sequencing

Genome sequencing is the same as DNA sequencing. Scientists use instruments connected to computers that process enormous amounts of data. They begin by melting the DNA to produce two strands. The scientist then adds a mixture of individual DNA nucleotides made of adenine, thymine, cytosine, and guanine. The scientist adds the enzyme DNA polymerase, flips a switch, and a new DNA molecule begins growing. The instrument keeps track of the nucleotides that jump onto the growing strand. By determining the order of nucleotide additions, the scientist automatically learns the nucleotide sequence in the original DNA.

Genomics has now become an essential part of these projects:

➤ Studying how different genes interact with each other to cause different outcomes in the body.

➤ Learning the ways that genes are regulated; signals that turn on genes and turn them off.

➤ Determining gene function (see the section on proteomics).

➤ Sequencing new species in which the genes have not yet been identified.

➤ Comparing genomes of different species.

➤ Finding errors in DNA replication or in gene sequences, and relating these errors to disease.

➤ Devising genes for use in gene therapy in disease treatment.

Proteomics

Scientists learned how to sequence entire chromosomes much faster than anyone could have predicted. Since 1995, the exact nucleotide sequence of almost 200 species' genomes are now known. Of course, this number represents a miniscule fraction of the total number of species on Earth. Genomics has a healthy future.

The successes of genomics has led scientists to think about biology's next big hurdle: learning the amino acid sequence of proteins. By doing this, scientists will begin to relate certain amino acid sequences with specific actions carried out by proteins. Welcome to proteomics!

Proteins may prove to be a much higher hurdle than gene sequencing turned out to be. Many more proteins than genes exist in every cell. The full complement of proteins of a cell is called its proteome. At this time, the technology for swiftly determining the amino acid sequence of a cell's proteome lags behind DNA sequencing technology.

The gap between genomics and proteomics is closing almost as fast as you can read this. The current hot areas of study in proteomics are adding a richer layer of understanding to our knowledge of how cells work:

Biology Tidbits

Gene Therapy

Gene therapy is a method of treating DNA-based diseases by delivering a gene directly into a person's cells. Scientists prepare a gene therapy drug by putting a desirable gene into a virus called a retrovirus. A doctor then injects the retrovirus into a patient's bone marrow, and the virus infects cells in marrow called stem cells. By this process, the desirable gene can either replace a faulty gene in a person's DNA or supply a missing gene.

The success of gene therapy depends on two features. First, retroviruses are good at inserting their genetic material into an infected cell's DNA. Second, stem cells give rise to all other cells of the body. By targeting stem cells, scientists have an efficient way of getting a desirable gene to the body tissue where it is most needed.

Gene therapy has seen limited use so far. Certain types of cancers that result from genetic errors now receive the most interest for gene therapy.

> ➤ Cataloging the proteins associated with human genes.

> ➤ Determining the functions of specific gene-protein combinations.

> ➤ Adding to our current knowledge of how certain proteins regulate specific genes.

> ➤ Defining complex protein networks.

> ➤ Further defining the basis of genetic variation between individuals and species.

Like DNA sequencing, proteomics depends on sophisticated instruments and a healthy dose of mathematics. Computer programs allowed scientists to decipher the human genome almost a decade before anyone thought possible. Computers that handle massive amounts of data will be as crucial to scientists tackling the human proteome.

Conservation Biology

Genomics and proteomics have been applied to two major aspects of biology: medicine and biodiversity. Conservation biology is a new science that applies genomics and proteomics to study biodiversity.

How can knowing the DNA sequence of a frog possibly help biologists slow biodiversity loss? Conservation biologists apply genomics to their work to accomplish the following:

> ➤ Determine genetic variation in species. A large amount of genetic variation indicates a species has a good chance of survival; a small amount of genetic variation indicates a species may be heading toward extinction.

> ➤ Identify specific genes associated with poor survivability, such as genes associated with disease, poor adaptations to the environment, digestive disorders, or breeding problems.

> ➤ Identify specific genes associated with robust populations. Biologists are interested in genes that improve resistance to infection, help an animal avoid predators, or enable a species to withstand changes in climate or exposure to pollution.

By analyzing species' genomes, conservation biologists get an idea of how closely species are related. Perhaps the day will come when biologists must choose to save one species at the expense of another that will go extinct. Their decisions will be helped by knowing the relationship between two species.

The principles of genomics have already been used to catch poachers who kill endangered species. The same DNA analysis used on television to catch criminals has been used to catch poachers. DNA from the blood of a poached animal usually contaminates clothing and vehicles. DNA analysis has connected perpetrators to particular kills and has even traced animal products such as ivory to an individual elephant herd.

The future of biology starts with the cell but now extends into global populations of plants and animals.

CHAPTER 33

The New Frontiers In Disease Fighting

In This Chapter

➤ How new technologies are developed

➤ Today's most promising new technologies

➤ An introduction to gene therapy

In this chapter you will learn how advances in our knowledge of biology are being used to find new ways to prevent and treat disease. The days are ending in which a doctor prescribes a drug that poisons diseased parts of the body. Left unsaid in this method of disease fighting is the hope that the poison works before it kills the patient along with the disease!

The newest technologies in disease fighting target very specific aspects of human biology. These technologies often use treatments called biologics, which you will learn about in this chapter. New disease fighting also makes use of electronic devices that monitor health and track the progress of treatments.

Doctors will increasingly combine emerging specialties within biology, such as nanobiology, with new and sophisticated medical devices. The new technologies in disease fighting are progressing quickly, but they are still known as cutting-edge technologies. Despite their promise, the vast majority of the world's patients continue to use traditional drugs and therapies to get well. This chapter will introduce some of the most promising technologies with the greatest chance of success worldwide.

How New Technologies Are Developed

New technologies for disease fighting come from laboratory experiments run by pharmaceutical and biotechnology companies or at universities. In commercial laboratories in the pharmaceutical industry, the laboratories fall under the heading of Research and Development, or R & D. Commercial R & D can take several years to invent a new technology. R & D is only the first step toward new types of disease fighting. Several more steps await the promising new technology before a government agency will allow it to be used to treat or prevent disease.

In the U.S. the Food and Drug Administration (FDA) oversees the testing and manufacture of drugs and medical devices for human use. A branch within theFDA called the Center for Veterinary Medicine has a similar role for the new drugs used in pets, companion animals, and food-producing animals.

After researchers design a prototype of the new technology with potential for fighting disease, the new drug or device enters FDA-approved testing for safety and effectiveness. Some of the first tests are in laboratory animals. The purpose of animal testing is to show

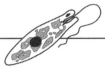

Inside a Cell

Nanobiology

Nanobiology is a new specialty in biology in which scientists study parts of the body as small as one billionth of a meter. This unit of measurement is called a nanometer and it equals 0.000000001 meter, or one-billionth of a meter. Put another way, there are 25,400,000 nanometers to an inch.

Nanobiology is used to study the smallest features of living cells. In disease research, nanobiologists study things like the connection point between a cellular molecule and a drug or the tiny pore used by a drug to slip inside a single cell. Nanobiology will be used to find errors in cell structure, mistakes in DNA that lead to disease, or foul-ups in the binding of a drug to a cell surface. By looking at these fine features of disease and disease treatment, biologists can learn more about disease mechanisms and improve drug design.

that the new technology is safe enough to be tested further in humans. This human testing is known as a clinical trial. Depending on the type of technology, a clinical trial may contain dozens to thousands of human subjects who all volunteer to participate in the *trial*.

After the clinical trial is finished, the inventor of a new technology fills out numerous forms and applications that explain the technology's role in fighting disease. If the invention is a drug, the FDA carefully reviews all the data that tell how the body absorbs the drug, how it is distributed in the body's tissues, how the drug is broken down or metabolized, and even how much is excreted and by what means, i.e., urine, feces, sweat, exhalation, etc. This information is part of two specialized branches of biology called pharmacology and pharmacokinetics. The reviewers also look for evidence of the drug's toxicology or potential to cause harm to healthy cells in the body.

Biology Tidbits

Clinical Trials

Clinical trials are studies using human subjects who volunteer to help medical research learn more about new drugs, medical devices, diet, medical procedures, or lifestyle behaviors. Every clinical trial is headed by a central authority, called a principal investigator, who is often also a medical doctor. A team of medical professionals helps run each trial which can last from a few weeks to several years and include dozens to thousands of subjects. The team collects data on the subjects' response to the treatment. Huge amounts of data are collected, checked for accuracy, and entered into computer databases. Statisticians analyze the data with a special type of mathematics called statistics. The principal investigator then publishes a report to summarize the clinical trial's results and conclusion. The FDA reviews the report to determine if the new medical treatment is safe for the general public and effective enough to be sold as a new product.

Clinical trials are also key to uncovering any bad side effects caused by the new medical treatment. The medical industry calls these side effects adverse events. Adverse events range from minor problems such as a skin rash to critical problems such as death. Thus, clinical trials act as important safeguards to the public.

Medical devices are also tested and the data are reviewed by the FDA. For devices, the FDA looks at the device's safety record, the manufacturing method, consistency (so that all newly made devices are of the same intended shape and size), and the quality of the raw materials and components that go into the manufacture of the device.

Biotechnology

Fighting disease used to be done by taking a pill or lopping off a limb. Today, we call the pill a drug and it is developed by an industry called pharmaceuticals. Fortunately, limb-cutting is not as prevalent today as it once was, but it nonetheless remains a way to treat certain diseases.

Biology Vocabulary

Pharmacology and Pharmacokinetics

Pharmacology is the study of how a drug affects systems, tissues, or cells in the body. Today's technology is able to investigate how a drug acts on specific molecules inside a person's individual cells.

Pharmacokinetics is the study of what happens to a drug once it enters the body. It starts from the time the drug is absorbed into the skin or into the bloodstream and ends when the body excretes the drug. Pharmacokinetics, or PK to biologists, includes the ways in which a capsulated drug exits the capsule, specific ways drugs are broken down inside organs, how they bind to cells, and how the drug reacts with other chemicals in the body.

Biologists sometimes say, "Pharmacology is what a drug does to the body and pharmacokinetics is what the body does to a drug."

Specialized fields in biology began to grow in the 1980s. These fields were based on science's ability to pull apart cells, extract the cell's DNA, and isolate specific genes in the DNA. When biologists learned how to put a piece of DNA from one organism into the DNA of another organism, genetic engineering was born. The method of moving genes around between different organisms became known as recombinant gene technology. (Bacteria were the first test model but this technology is now used in higher plants and animals.) This term was coined because biologists literally recombined genes in ways nature never did.

Biotechnology makes use of genetic engineering and recombinant gene technology to make new products to benefit humans and the environment. A branch of biotechnology that focuses only on new drugs made from recombinant gene technology is called biopharmaceuticals. Biopharmaceuticals is a fast-growing industry with more than 300 products on the market and many more currently being tested in laboratories or in clinical trials.

Even after all the laboratory testing and clinical trials are complete, the FDA carefully watches the manufacturing of new drugs and devices. Manufacturing is a key point in assuring that the new product is completely free of contamination from bacteria, fungi, or non-living contaminants such as dust, hair, and dirt.

Today's New Technologies in Disease Fighting

The new technologies for fighting disease fall into two main categories: (a) prevention and (b) treatment. We could also classify them by whether the technology is completely biological, completely non-biological, or a combination of both biology and a device.

Meet 10 of the most promising technologies in disease fighting.

➤ **Biologics**: Short for biological products, these are natural materials made from cells, cell parts, or tissue. They are promising as treatments for many diseases because they are intended to mimic the action of the body's tissue. Some biologics are used when no other medical treatment is available, but some, such as vaccines, have been used throughout the world for decades. (Read more about biologics in the next section.)

➤ **Gene therapy:** This procedure is a type of disease treatment or prevention in which genes in a patient's DNA are replaced with new genes. A later section in this chapter provides more details on gene therapy.

➤ **Genetic testing:** This technology is currently being used as a tool for checking to see if a person is at risk of developing a specific disease. By analyzing a sample for the presence of certain genes, proteins, or enzymes, a doctor gets a clearer picture of an individual's genetic makeup and possible future health issues. (The proteins and enzymes studied in genetic testing are called genetic markers.) Genetic testing is also available for pregnant women for determining the presence of certain inherited diseases or abnormalities in the fetus.

➤ **Nanotechnology:** Nanotechnology is the study or manipulation of entities so small they are measured in nanometers. When nanotechnology focuses on biological things, the field is called nanobiology. Nanobiotechnology is the commercial development of nano-sized biological materials for use in applications such as medicine, cosmetics, and the environment. Typical materials being studied in nanobiology are molecules and even atoms.

➤ **Tissue engineering:** This technology goes by a few different names. It involves the ability to artificially grow a human organ, such as a lung, in the laboratory. The fully formed organ can then be implanted into a patient. The technology gives hope to people waiting for an organ transplant. Tissue engineering begins with stem cells, which serve as the building blocks for new organ growth.

➤ **Electronic aspirin:** This is a nickname for electronic devices that can be implanted in the body to stimulate or block specific nerve transmissions. Devices like this are already being tested for pain management.

➤ **Genome editing:** Also known as CRISPR, this emerging technology finds defective genes or mutations, and then corrects them. The technology is adapted from bacteria, which routinely fix their own DNA when it gets injured. (CRISPR stands for clustered regularly interspaced short palindromic repeats, a term only geneticists could love.)

➤ **Brain computer interfaces**: Known as BCI, this technology uses headsets to monitor brainwaves of a person, such as a paraplegic. By concentrating on a specific body movement, the technology operates a part of the body, such as a limb. This technology is far away from practical use today, but is being investigated as an aid to Alzheimer's patients and possibly for pain management.

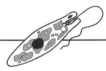

Inside a Cell

Stem Cells

Stem cells are found throughout the human body. Their role is to act as a starting point for producing new specialized cells and tissues when the body is growing or repairing itself. Sometimes adult stem cells divide to simply make more stem cells. But often they divide and then accrue characteristics of the new tissue they will become. This is called differentiation.

Human embryos begin as stem cells. During embryonic development, the cells differentiate into the specialized cells, tissues, and whole organs that will make up the fetus and carry out specialized functions for a lifetime.

Stem cell research seeks to use embryonic stem cells to cure or prevent disease. Researchers plan to someday use stem cells in regenerative medicine. In this type of medicine, stem cells may be valuable for the following uses: to bolster immune systems lacking sufficient immune cells, to repair burn injuries, to correct birth defects, to improve arthritic joints, to cure macular degeneration, and to treat spinal cord injuries.

➤ **Bioinformatics:** This technology uses the power of computers to analyze massive amounts of biological data. These analyses can spot relationships within biology that were previously very difficult to uncover. For instance, bioinformatics may be able in the future to find relationships between genes and disease, environmental chemicals and cancer, drugs and side effects, and perhaps inherited traits and behavior.

➤ **Lab-on-a-chip:** This offshoot of nanotechnology creates small discs about the size of a half dollar or smaller. The discs called wafers or chips contain hundreds of tiny hotspots filled with enzymes, antibodies, and other biological things that tend to react with the body's cells, blood, and fluids. An electronic circuit inside the chip records the biochemical reactions and gives a signal when the test is done. This convenient technique can diagnose infection, disease, or find other abnormalities in the body. These chips are especially promising for doctors who must go to remote places to diagnose illness in people who have no medical care available to them.

Biology Vocabulary

Allergens

Allergens are extracts from natural things that are notorious for causing allergies. The most common allergens used in medicine today are: pollen, mold, insect parts, venom, dander, and food proteins. Putting the extracts of these allergens into or on the body elicits an allergic reaction in susceptible people. This is a common way to diagnose allergy.

Medical professionals follow one of three procedures for exposing a person to an allergen. The first method is injection into the bloodstream. The second method is injection under the skin. The third method involves applying several different allergens onto different locations on the skin (usually on the back) in a procedure called the patch test. The allergen is soaked into a patch which is then taped onto the patient's skin for one to a few days. A positive reaction at specific application sites can help diagnose specific allergies.

In allergenics, allergen extracts in tablet form may also be placed under the tongue. This method is used for building a person's immune response to the allergen and thus can treat some chronic allergies.

Biologics

Biologics are products of biological origin used to treat medical conditions. The main biologics in use today or being tested for imminent use in humans are: vaccines, blood, blood components, cells and tissues, stem cells, genes, proteins made by recombinant gene technology, and allergens. Allergenics is a field of medicine that uses natural substances that normally cause an allergic reaction in people. If used a certain way allergens can help diagnose or even prevent allergy.

Biopharmaceutical companies are the main producers of biologics. Biologics are now common in diagnostic kits, disease treatments, and disease prevention. Biologics are difficult to produce because of the high risk of contamination. Producing these substances under sterile conditions is of utmost importance in order to protect the health of the end-user, the patient. In addition, many biologics are hard to keep stable for long periods. In general, biologics tend to be made in injectable form rather than as a pill. They are strictly regulated by the FDA.

Gene Therapy

Gene therapy is part of a larger field in biology called genomics. Because it uses genes as its main form of treatment, it is also part of biologics.

In gene therapy, a treatment might be able to correct a medical condition at its source, the gene. Some diseases are caused by a mutation in a gene or by a missing gene. In both cases gene therapy will be useful in delivering the correct gene to its place in a person's DNA. Once the correct gene is in place, the body can be expected to carry out its normal functions. The obvious advantage to gene therapy is that it avoids drugs and surgery.

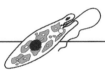

Inside a Cell

How to Get a Gene into a Cell

Gene therapy's first challenge is to get a gene into a patient's existing DNA. A gene that can cure or prevent disease does no good unless it can be incorporated into a person's DNA. The most common method of delivering new genes into cells is by exploiting viruses. Viruses are, after all, experts at getting into your cells and inserting their genes into your DNA. Gene therapy now uses a variety of viruses for this delivery. Virologists first strip the virus of its disease-causing ability before loading it up with the gene to be delivered. The main types of viruses used in gene therapy are adenoviruses, retroviruses, and Herpes Simplex Virus. The viruses used in gene therapy are collectively called viral vectors.

When viruses do not work for specific types of gene therapy, scientists can also use small, fatty globules called liposomes. Sometimes a strand of DNA can be used all by itself; biologists call it naked DNA. Naked DNA is much less effective than viruses or liposomes for delivering genes into cells.

Gene therapy is far away from being a standard medical approach to disease. The technology is difficult to master. The manufacture of genes to be put into the body must eliminate all contamination. Furthermore, a scientist must find ways to ensure that the new gene will remain active once it is inside the body.

So far, gene therapy has made inroads against diseases that have few options for cure. It is now being used in some hereditary diseases such as rare immune disorders, hemophilia, the blood disease beta-Thalassemia (a defect in the oxygen-carrying protein), hereditary blindness, and Parkinson's disease.

Which diseases will new technologies target first?

Some new technologies in disease fighting can be applied to a wide variety of diseases. For example, the biologic products called vaccines prevent a large number of different infectious diseases. Other types of biologics can be used for diagnosing and preventing additional diseases that afflict every organ and system of the body.

Other technologies are very specialized and were developed for rare health conditions. Before long, researchers see the potential of expanding that same technology into new areas in medicine. For example, electronic aspirin began as an approach for alleviating chronic pain. Technology companies have quickly adapted the idea to new applications. Electronic devices that were invented for pain management are now being tested as treatments for stroke and hearing ailments.

The diseases that will be first to benefit from new technologies are limited only by your imagination. In general, you can be confident that the following diseases or disease groups are high on everyone's list of priorities: heart disease, stroke, cancer, diabetes, cystic fibrosis, and Alzheimer's disease.

We must be realistic and remember that the drug and medical device industry is a business. Diseases that affect the most people get the most attention in research, clinical trials, and the rollout of new products. But small companies and startups often introduce the most innovative technology. These innovations sometimes target medical conditions that affect a small number of people in the population or the diseases most difficult to treat.

The ability of technology to reveal the secrets of living cells gives hope to all people who depend on new treatments for fighting disease.

CHAPTER 34

How To Understand Biology In The News

In This Chapter

- ➤ Understanding how biologists talk
- ➤ Biology is an inexact science
- ➤ An introduction to statistics used by biologists
- ➤ How to spot junk science

In this chapter you will learn to make better sense of news from the world of biology, especially when such news sparks debate. Can biology be controversial? Can it stir politicians, preachers, and the public into heated arguments? Yes, and new discoveries in biology can also cause the most heated debates among biologists themselves. When trained scientists engage in public arguments, it is hard for non-scientists to figure out where the truth lies.

Consider these questions to dispel any ideas that biology is a dry subject archived in dusty textbooks or sealed inside a jar filled with pickled tadpoles.

- ➤ Do vaccines endanger or improve health?
- ➤ Is evolution theory or fact?

➤ Is cancer curable?

➤ How much do humans affect climate change?

Indeed, is climate change a biological phenomenon, a physical result of industrialization, or both? Furthermore, when hearing today's the debates on climate change, you might wonder if climate change is a discovery based on sound science or the invention of politicians, environmental activists, or fiction writers.

Biology Tidbits

Peer Review

Peer review is a process in which a small group of scientists critique the work of another scientist who has completed a study and written an article he wants to publish in a particular scientific journal. The journal's editor selects the team (usually three people) of reviewers with expertise in the scientist's specialty. The reviewers examine the article's description of the methods, data, analyses, and conclusions and send the author anonymous feedback. The author then makes any requested changes to the article or rebuts the reviewers' critique. In the end, the journal publishes a fully vetted article that presents new data and findings in a particular scientific field.

The modern form of scientific study and critique was first proposed by English statesman Francis Bacon in the 17th century. The Royal Society of London refined the technique and peer review was eventually adopted by almost every scientific society in existence today. Peer review has since evolved from a closed system dependent on experts to a more open forum of online communities. Though rare, editor-selected reviewers could impart their personal biases to their reviews. The online expansion of information flow has led many scientists to believe that traditional peer review has become antiquated. Rather than rely on a tiny team of reviewers to determine our distribution of knowledge, scientists are increasingly seeking an open forum for review and publication. In this way, scientific discoveries can be critiqued by trained scientists as well as amateur scientists and non-scientists. This type of public review represents a small, but growing, portion of all published science. In time, we will learn the benefits and perhaps some drawbacks of public review of complex scientific ideas.

Biology news can be difficult to interpret for three main reasons: (a) the biological world is uncertain, (b) biology is inexact, and (c) biologists often come up with more than one answer to many of the main questions in biology.

Many of our most pressing health and environmental issues are a blend of rigorous studies, unbiased scientific analysis, and peer-reviewed reporting. Into this mix society adds the news media, political agendas, religious beliefs, and public input via social media. A harried scientist emerges from his laboratory to announce a promising, though untested, theory that might save lives in the next decade. A small sampling of the scientist's data enters the public domain where it is dissected, discussed, argued, and sometimes distorted and misused. Before the untested theory has a chance to fly, many non-scientists and even trained scientists may have already shot it down.

It takes a strong person to persevere and continue his studies in the face of criticism. Fortunately, the fortitude of scientists named Kepler, Copernicus, Galileo, and hundreds of others to the modern day have helped explain our universe even though they were forced to overcome condemnation from their peers.

This chapter explains why biology news can be hard to interpret and understand. It shows you how to read between the lines when scientists explain their discoveries. You will learn about the common terms used by scientists and understand why professional biologists explain things in so confounding a manner! This chapter also helps you make sense of the breathless reporting on the most astounding discoveries in biology. After reading this chapter, you will gain a better handle on very complex information that emerges from research laboratories, university press conferences, and yes, even from the mouths of celebrities and politicians.

How Scientists Talk

We face uncertainty every day in almost everything we do. Despite this, many people expect scientists to be certain of their facts all the time. This is unrealistic because the role of the scientist is to explore the world's uncertainties. Once they find answers to our questions scientists delve into new uncertainties. The pursuit of the unknown comes with a few startling successes but also many failures. When an experiment fails or gives a completely unexpected outcome, a non-scientist might assume that science has lost its way. On the contrary, a scientist knows that the unexpected might mean a discovery is right around the corner.

Biologists are scientists. Scientists do not know everything and science cannot yet explain everything in our universe. For this reason scientists speak in language that tries to convey the uncertainty and inexactness of our biological world. By no means does this imply that scientists are confused, befuddled, or on a devious mission to mislead the public. University of Michigan professor Henry Pollack once said, "When scientists acknowledge that they do not know everything about a complex natural phenomenon such as the spread of disease through an ecosystem, the public sometimes translates that to mean that scientists do not know anything about the subject." Don't make that mistake.

Biologists often use catchphrases to explain new, and sometimes complex, discoveries. By using these phrases they remind us of biology's uncertainties:

➤ "These results suggest..."

➤ "The data support our hypothesis that..."

➤ "...approximately..."

➤ "...with a probability of..."

➤ "...is now proposed as the cause of..."

➤ "Our study provides evidence that..."

➤ "It is widely recognized that..."

Good grief! No wonder biologists never sound like they are sure of anything! The public tends to doubt a scientist's credibility when they hear these kinds of qualifier phrases. But biologists know that as soon as they state anything unequivocally, new information may soon prove them wrong.

In chemistry, combining two hydrogens with one oxygen always produces a water molecule. In physics, dropping an apple from a tree always sees the apple plunge to earth. Gravity never disappoints. But in biology, exceptions to every rule flourish. No one human is identical to any other human; even identical twins differ a little. No bacterial cell grows exactly at the same rate as its neighbor; a herd of elk in Montana graze differently than a herd in Canada; a flu vaccine works well for Mary yet causes a sore throat in her brother Bob.

To understand biology better, remember that the universe is uncertain, biology is inexact, and life can be inconsistent.

Knowing a Little Statistics Can't Hurt

A biologist might say, "Nothing in biology is black and white. Everything is a shade of gray." That is a poetic way of restating biology's tendency toward uncertainty and inexactness. In biology experiments, scientists see a range of outcomes rather than one single definitive outcome. This is because biology is made up of populations of living things. These populations tend to possess ranges of characteristics rather than having a single characteristic.

Biology is made of as many different types of populations as you can imagine. If you were to draw a dot on a paper for each member of a population, a picture would begin to emerge.

More dots cluster toward the midpoint of the population and decreasing numbers of dots accumulate toward the edges. The final picture looks like a slightly spread-out diagram of the Liberty Bell. Thus, scientists refer to this image as a bell curve. You can draw a bell curve for any population in biology that you can dream up: the running speed of 1,000 gazelles; the intelligence levels of gray whales; the hearing acumen of barn owls; the test scores of high school Seniors in Kalamazoo, Michigan; you get the idea.

Biologists use statistics to make sense of populations and what the data from populations are trying to tell us. Statistics is a branch of mathematics. Statistics begins with very simple calculations such as figuring out the average (called a mean) value in a population of values. For example, let's assume that the mean test score for those students in Kalamazoo is 75. On a fairly symmetrical bell curve, about half of all students score higher than 75 and about half score lower.

Statistics become more and more complex as scientists try to learn more about the sometimes unwieldy populations they study. They discuss their findings using terms such as "mean," "probability," and "standard deviation." When you hear these terms, you do not have to be a math expert to decipher the message. These terms tell you that a biology study gave a range of results. The scientist then analyzed the results using statistics to make sense of the data and draw a conclusion about the study.

Here are reasons why biologists use some common statistical terms:

➤ Population: To show that results from a particular study cover a range of outcomes rather than a single answer to a question. Example: all the individual sequoia trees in Sequoia National Park.

➤ Mean: To tell us where most of the individuals exist in a population. Example: the batting average of the New York Yankees.

➤ Median: To show where in a population lies the exact point in which half the population is below and half is above. Example: the midpoint sale price of all houses in Somerset County, New Jersey.

➤ Standard deviation: To estimate how far from the mean an individual exists in a population. Example: an obese 40-year-old man is two standard deviations from the mean body weight of all 40-year-old men in North America.

➤ Probability: To estimate the likelihood that a specific outcome will occur. Example: the probability of rain in the Amazon Basin is high; the probability of rain in the Sahara Desert is low.

➤ Variability: To indicate the amount of inexactness present in a set of measurements. Example: the shapes of snowflakes.

Biology Vocabulary

Probability

The term probability in statistics is short for "the probability that a certain event will occur." For example, the probability of flipping heads on a quarter is one in two. The probability of rolling a five on a die is one in six. The probability of you winning the lottery (assuming you actually buy a ticket) might be about one in 20 million. Biologists use probability to help predict what may happen in the future based on the events that have occurred in the past. They learn about the past by studying various characteristics of populations. Thus probability helps doctors predict which drug will be most successful against an infection. It helps climate change experts predict ocean temperatures in the next decade. Probability also helps biologists study events as simple as when a rose will bloom to a question as complex as when a species might go extinct.

How to Spot Junk Science

As part of human nature, we often believe what we want to believe. To do this we ignore facts as long as we possibly can to confirm ingrained beliefs. Consider some of the following sentences that have been uttered in the course of history:

➤ "The Earth is the center of the universe."

➤ "The world is flat."

➤ "She is a witch."

➤ "AIDS is a punishment from God."

At the time these thoughts were said, the speaker was convinced of their truth. Scientists' perseverance in finding the truth won out. Now think about more modern discussions in biology:

➤ "Wolves are a dangerous threat to humans."

➤ "Vaccines harm children."

➤ "Climate change is a myth."

➤ "Evolution cannot be proved."

➤ "He loves me."

Although many people hold these beliefs, scientists have accrued massive amounts of data to uncover the truth about wolves, vaccines, global warming, evolution, and numerous other issues in the biological world. In time, the truth always emerges (except maybe for that last one).

Junk science, or pseudoscience, occurs when people use scientific data for their own purposes. For example, a politician supported by the coal industry may distort certain environmental data to show that coal mining has no effect on the environment. The politician may accuse environmentalists of using junk science to show that coal mining hurts ecosystems. Meanwhile, the environmentalists accuse the politician of applying junk science to support mining. Sometimes, junk science is in the eye of the beholder, but sometimes it is just plain junk.

Signs of junk science:

> ➤ Results that cannot be supported by any other study.

> ➤ Conclusions drawn without any data.

> ➤ Data that were not analyzed with statistics or checked for accuracy.

> ➤ Results selected to support a belief (called cherry-picking) while ignoring a larger amount of data that refutes that viewpoint.

> ➤ Illogical conclusions drawn from accurate data.

> ➤ Studies touted by politicians, religious leaders, or other spokespeople, such as celebrities, but no scientific experts.

> ➤ Studies touted by scientists who are paid by companies to give a certain opinion, i.e., the expert witness.

> ➤ Discoveries that appear only on the cover of tabloids or in social media.

> ➤ Discoveries linked to a new product but with no support or mention in scientific articles or university research press releases.

> ➤ Data, results, or studies that were entirely made from someone's imagination (called dry-labbing).

It is OK to view any scientific news with skepticism until more facts accumulate. Over time, you will realize that the truth in biology is supported by many dozens to hundreds of studies by various research groups. A consensus forms and old beliefs begin to give way to new, more accurate information. Still, people continue to believe what they want to believe.

Sometimes, our beliefs are harmless and even a fun way to look at biology. I see no reason why we cannot have fun debates about the following:

> ➤ UFOs.

> ➤ Bigfoot.

> ➤ The Loch Ness monster.

> ➤ Ghosts.

When pursuing controversial subjects in biology, look for websites and publications from a diverse variety of research universities and scientific organizations. Be aware of the pitfalls that lead many non-scientists, and even trained scientists, to mistake the facts of biology. Think about these things when you are unsure of what to believe:

> ➤ What are the facts compared to what I want to believe?

> ➤ Is a personal bias or fear leading me to an illogical conclusion?

> ➤ Is my religion telling me to believe something I know is untrue?

> ➤ Is a politician asking me to ignore the preponderance of evidence in favor of a political platform that denies the evidence?

> ➤ Am I making a conclusion based on emotion rather than fact?

Don't immediately believe a study is junk science because someone else says it is. Seemingly farfetched ideas have proven over time to be correct. Scientists who weather blistering attacks in the press and in their own scientific circles have been proven right, though it may have taken years or decades for the truth to win out. Many pathogens, diseases, and idiosyncrasies of DNA are known today because of the steadfastness of their discoverers.

Biology Tidbits

The Very Small and the Very Big

Humans find difficulty in imagining things on a scale much different from where we live. Although it is easy to understand what a flea is and an elephant and even a massive blue whale, modern biology has moved off this human scale. Nanobiology is the study of biology on the scale of atoms, which are the smallest parts of a single molecule. We have a hard time visualizing how a biologist can possibly study a speck only a billionth of a meter long. Conversely, concepts such as climate change, global warming, extinction, and even human-induced earthquakes are difficult to understand. The world seems so huge, most people do not believe their actions can impact the planet. Of course, if several billion people all take the same action, such as burning fossil fuels, their impact on the Earth can be evident. Ecology and environmental science look at biology on an Earth scale. The field of astrobiology delves into life that originates in places beyond Earth.

This type of dedication will lead biologists to someday find the cure for cancer, discover life in the deep earth, find new species at the ocean floor, or decide if Martians exist.

When thinking about the credibility of science in the news, keep in mind that mistakes happen. An ethical scientist who has discovered he has published an article containing a big mistake will immediately contact the journal's editor. The journal then publishes a retraction. If the mistake is newsworthy, the public might start to think that scientists don't know what they are doing. Covering up mistakes is never acceptable, but a full retraction of results that are in error is all part of the difficult job of conducting research.

Our Perception of Biology

The uncertainty within biology leaves a door open for opinion. We expect trained biologists to develop opinions based on their education and experience in their field of study. Amateur scientists do much the same to fill in knowledge gaps by reading technical literature and talking to scientific experts. Non-scientists, however, might develop opinions based more on personal experience than science background.

Everyone from trained scientists to non-scientists develops opinions about biology based on long-held perceptions of our place in the living world. Where do you fit in the following categories posed by ecologists to describe our relationship to nature?

> ➤ Biology has instrumental value. This means the Earth is a resource for humans to use as they see fit. After human needs are met, people can focus on helping other species and the environment.

> ➤ Biology has intrinsic value. This means the Earth does not exist solely for human use but plays a bigger, as yet undefined, role in the larger universe.

> ➤ Biology is part of ecological economics. This doctrine states that a true value can be put on everything in the biological world.

> ➤ Biology is part of neoclassical economics. By this theory, all things in the biological world are here for humans, and when a resource is used up technology will find substitutes.

> ➤ Biology is part of environmental economics. This approach teaches that some things in biology are irreplaceable so that humans must find ways to live sustainably within the biological world.

So now you see that biology can be part of philosophy as well as science. Your understanding and perceptions of biology may fall somewhere on a wide spectrum. At one end resides the belief that humanity has divine reign over all other living things. At the opposite end of the spectrum is the belief that humans play a small role in an enormous universe of life. Some of the truths in biology can be very difficult to find.

APPENDIX A

Glossary

Active site—The location on an enzyme that connects with a substrate.

Adenosine—One of four nucleotides that make up DNA; also part of the energy compound ATP.

Allergen—Any substance that enters the body and causes a response from the immune system and is attacked by antibodies.

Allergic reaction—The binding of a specific antibody with an allergen.

Ammonia—A nitrogen-containing compound (NH_3) that is part of nitrogen metabolism in microbes and plants.

Antimicrobial—Pertaining to any substance that inhibits the growth of microbes.

Archaea—Bacteria-like microbes that are one of two types of prokaryotes.

Biochemistry—The science of chemical reactions taking place inside a living cell or organism.

Biodiversity—Number and variety of species in a particular region of the Earth.

Biota—All the living things on Earth.

Bloom—A sudden burst of growth of an organism due to an influx of nutrients into its environment.

Brackish—Water that is slightly salty but not as salty as the ocean.

Carnivore—A type of animal that evolved to eat other animals.

Carotenoid—A type of pigment that participates in photosynthesis.

Cell plate—A wall that forms down the middle of a dividing plant cell.

Chemoreceptor—A substance on a cell surface that detects the presence of a specific chemical in the environment.

Chitin—A strong polymer that protects the exoskeleton of insects.

Chromosome—All the genetic material contained in a cell.

Codon—A sequence of three DNA bases that corresponds to an amino acid.

Colony—Any group of many microbial cells of the same species.

Continental shelf—The underwater perimeter of each continent on Earth that extends into the ocean.

Cytoplasm—The watery material that fills most of a cell's contents.

Deciduous tree—A tree that loses its leaves each winter and grows new leaves in the spring.

Domain—The main division of living things; Earth's biota is divided into three domains.

Dormant—A state of a cell in which little or no metabolism takes place for an extended period.

Electron—A negatively charged component of an atom.

Electron transport chain—A series of steps carried out during respiration in which energy is generated and oxygen consumed.

Enzyme—A protein that facilitates chemical reactions in cells.

Eukaryote—A type of cell characterized by membrane-bound organelles.

Extremophile—A microbe that survives in extreme environments of high heat, cold, dryness, high salt, etc.

Fermentation—An energy-generating pathway that consumes organic compounds rather than oxygen and produces a variety of different organic compounds.

Fertilization—The combining of two haploid reproductive cells to form a diploid cell.

Food chain—A relationship of living things in which an animal is preyed on by another animal, and that animal is then eaten by a larger animal.

Functional group—The part of a molecule that interacts with another molecule in a chemical reaction.

Gene expression—The conversion of the information held in a gene to a functioning protein.

Genetic code—The lineup of nucleotide bases of DNA that corresponds to functioning proteins and an organism's physiology.

Global warming—The gradual increase in the Earth's average atmospheric and surface temperatures.

Globulin—A type of protein; globulins make up antibodies.

Glycogen—A large molecule made of repeating glucose units and used by cells for storing energy and carbon.

Glycolysis—The main set of steps in metabolism by which organisms get energy from glucose.

Gradient—Any difference in a molecule's concentration from one side of a membrane to the opposite side.

Herbivore—A type of animal that evolved to eat only plants.

High-energy phosphate bond—A chemical connection linking phosphate (PO_4) to adenosine.

Homologous chromosome—Chromosome pairs of the same length, one inherited from the father and one from the mother.

Hydrothermal vent—A fissure at the sea bottom that releases superheated gases from the Earth's mantle.

Imbalance—Condition in which an ecosystem contains too many of some species and too few of others.

Infectious disease—A disease caused by a microbe.

Integumentary system—Outer covering of a mammal's body composed of skin, hair, fur, nails, etc.

Lipid—A type of compound that is insoluble in water and includes fats.

Lymph—Colorless fluid that circulates in the lymphatic system; the lymphatic system functions to return cells and proteins to the blood.

Membrane—A thin layer that holds in a cell's contents and keeps it separate from the outer environment.

Metabolic pathway—Any series of connected steps that is part of an organism's energy and nutrient use.

Metabolism—The sum of an organism's chemical reactions that build large molecules from small ones and break down large molecules into smaller molecules.

Methane—A gas (CH_4) produced from carbon dioxide by certain microbes; also natural gas found in the earth.

Methanogen—A type of microbe that produces methane.

Microenvironment—Any specialized environment that requires its inhabitants to be adapted to it.

Microtubules—Filaments in a cell that support the chromosomes during mitosis and cell division.

Neutron—An uncharged component of an atom.

Nitrogen fixation—The activity of microbes in which the cell absorbns nitrogen from the atmosphere.

Nucleic acid—Refers to either DNA or RNA.

Nucleus—The organelle of all eukaryotic cells that contains the cell's DNA.

Optimal range—The environmental conditions in which an organism grows best, defined mainly by temperature, oxygen, sunlight, acidity, and nutrients.

Parasite—An organism that lives on or in another organism and causes harm to it.

Pathogen—A disease-causing organism.

Permeable—Pertaining to the ability to let certain substances move through a barrier.

Peptide—A strand of amino acids not long enough to be classified as a functional protein.

Peptidoglycan—A large molecule that strengthens the cell wall of bacteria.

Phylum (plural: phyla)—In the classification of living things, a category that fits into a kingdom above it and contains classes below it.

Plasmid—Pieces of circular prokaryotic DNA that occur in the cytoplasm separate from the main chromosome.

Polymer—Any large molecule arranged in a strand, either straight or branched.

Polypeptide—A long peptide made of at least a hundred amino acids.

Polysaccharide—A long strand of sugars.

Primary producer—An organism that performs photosynthesis, converting the Sun's energy to chemical energy for use by animals.

Prokaryote—A type of cell characterized by non-membrane-bound organelles and DNA exposed to the cell cytoplasm.

Protein—A strand of amino acids that performs a function in cells.

Proton—A positively charged component of an atom.

Reaction—The conversion of a chemical into another chemical.

Red tide—A sudden overgrowth of red algae in water.

Resin—A hydrocarbon produced by various trees, such as conifers.

Respiration—An energy-generating pathway that consumes oxygen and produces carbon dioxide.

Ribosome—The protein manufacturing organelle of cells.

Ruminant—An animal with a compartmentalized stomach and the ability to digest and get energy from plant fiber.

Sterol—A type of compound similar in structure to steroids and including cholesterol; important component in the membranes of archaea.

Symbiotic relationship—Also called symbiosis, any close relationship or contact between two unrelated organisms.

Template—A compound that serves as a blueprint for a protein to build another compound.

Trait—An inherited characteristic.

Transmission—The movement of a pathogen from the environment or another individual to a susceptible animal or plant; the conduction of an impulse along a nerve.

Tundra—A type of earth that contains limited plant growth due to a short growing season.

INDEX

A

Active site 273
Adenosine 273
Algae, Protists, and Fungi 155–163
  algae 156–157
    amoebae
      E. histolytica 158
      Entamoeba 158
    Charophyta 156
    Chlorophyta 156
    Chrysophyta 156
    Euglenophyta 156
    Harmful Algae Blooms (HABs) 157
    Phaeophyta 156
    Pyrrhophyta
      dinoflagellates 156
    Rhodophyta 156
    Volvox 157
  apicomplexans 158
  fungi 160–163
    ascomycetes
      Neurospora crassa 162
      sac fungi 162
    basidiomycetes 162
    commensalism 161
    filamentous growth 162
    heterotrophs 160
    molds 162
    multicellular fungi 160–161
      hyphae 160
      mycelium 160
      reproductive structure 160
    mutualism 161
    symbiotic relationships 160
    yeasts 163

    Saccharomyces 163
    S. cerevisiae 163
  lichens 161
  protists 157–159
    amoebae 158
    ciliates
      Paramecium 158
    flagellate 158
    protozoa 157
    slime molds 159
    sporozoa
      Plasmodium 158
Allergen 273
Allergic reaction 273
Ammonia 273
Animal Body: The Endocrine, Immune, and Sensory Systems 193–200
  endocrine system 194–196
    adrenal glands 194
      blood glucose 194
      corticoids 194
      epinephrine 194
      norepinephrine 194
    hormones 195
      reception 195
      response 195
      signal transduction 195
    hypothalamus
      pituitary gland 194
    pancreas 194
      glucagon 194
      insulin 194
    parathyroid gland
      parathyroid hormone 194
    pheromones 195–196
      aggregation pheromones 196

sex pheromones 196
pineal gland
  melatonin 194
pituitary gland 194–195
reproductive glands (gonads)
  testes (androgens) 194
Reproductive glands (gonads)
  ovaries (estrogen, progesterone) 194
thyroid gland
  calcitonin 194
immune system 196–199
  acquired immunity
    active acquired immunity 198
    passive acquired immunity 199
  antibodies 198
  bone marrow 197
  duality 197
  lymph nodes and vessels
    lymphocytes 197
  lymphocytes
    B cells 198
    T cells 198
  natural immunity 198
  Peyer's patches 197
  redundancy 197
  spleen 197
  thymus gland 197
kinetic energy 200
mechanical energy 200
potential energy 200
sensory system 199–200
  equilibrium 200
  mechanoreceptors 200
  sensory receptors 199–200
    chemoreceptors 199
    electromagnetic receptors 199
    mechanoreceptors 199
    pain receptors 199–200
    thermoreceptors 200
Animal Body: The Nervous System and Circulation
    183–191
action potential 186
body metabolism 187–188

ectotherms
  fish 187
endotherms
  warm-blooded animals 187
hibernation 188
homeostasis 187
thermoregulation 188
  adjusting metabolic rate 188
  circulatory system adaptations 188
  evaporation 188
  insulation 188
circulatory system 189–191
  blood 191
    erythrocytes (red blood cells) 191
    immune system cells 191
    leukocytes (white blood cells) 191
    lymphocytes 191
    platelets 191
  blood vessels 191
    arteries 191
    capillaries 191
    veins 191
  closed circulatory systems 190
    heart 190
  open circulatory systems 190
    gas exchange 190
nervous system 189
  chordate 189
  glia 189
  insect style 189
  nerve cords 189
  nerve nets 189
  neurons 189
  squid style 189
  star 189
resting potential 186
tissue structure and function 184–186
  connective tissue 186
    adipose (fat) tissue 186
    blood 186
    bone 186
    cartilage 186
    collagenous 186

elastic 186
  rectangular 186
epithelial tissue 184
  cell shapes 184
  glandular 184
  simple 184
  stratified 184
muscle tissue 185–186
  actin 185
  cardiac 185
  myosin 185
  skeletal 185
  smooth 185
nerve tissue 186
  neurons 186
Antimicrobial 273
Archaea 273

B

Basics of Animal Reproduction 201–206
  asexual reproduction in animals 202
    budding 202
    fission 202
    fragmentation 202
  breeding 205–206
    acrosomal reaction 205
    cortical reaction 205–206
    entry 205
    fusion 205
    sperm-egg contact 205
  fertilization 203
    external 203
    internal 203
  human female reproductive system 203
    mammary glands 203–204
    ovaries 203
    oviducts 203
      fallopian tubes 203
    uterus 203
    vagina 204
    vulva 204
  human male reproductive system 204
    bulbourethral glands 204

ejaculatory duct and urethra 204
epididymis 204
penis 204
prostate gland 204
scrotum and testis 204
seminal vesicles 204
  semen 204–205
vas deferens
  sperm cells 204
reproductive cycles 202
  breeding 202
  nonbreeding 202
  raising offspring 202
Biochemistry 273
Biodiversity 261–265, 273
  biodiversity hotspots 262–263
  biodiversity loss 262
  endemism 263
  habitat loss 262
  keystone species 264
  premature extinction 262
  protecting biodiversity 263
  threats to biodiversity 263–264
    climate change 264
    fuels
      global warming 264
    habitat fragmentation 263
    habitat loss 263
    nonnative species 263
    overpopulation
      urban growth 264
    pollution of ecosystems 263
    poverty 264
    protect habitat 263
    removal of keystone species 263
  value of biodiversity 265
    antibiotics 265
    anti-cancer drugs 265
    building materials 265
    chemicals 265
    clothing materials 265
    foods 265
    indicator species 265

Biology Tidbits 4, 5, 12, 24, 32, 33, 41, 60, 68, 73, 75, 76, 83, 90, 92, 114, 116, 118, 126, 130, 132, 133, 135, 139, 149, 157, 161, 167, 185, 188, 190, 195, 214, 218, 227, 231, 235, 236, 239, 250, 256, 271
Biology Vocabulary 6, 7, 13, 14, 16, 17, 29, 35, 38, 42, 48, 51, 59, 61, 66, 68, 69, 75, 78, 79, 80, 91, 96, 101, 108, 112, 113, 125, 126, 133, 134, 138, 150, 158, 160, 161, 177, 179, 187, 197, 200, 203, 205, 213, 215, 217, 220, 221, 223, 227, 228, 234, 246, 247, 251, 254, 258, 259, 262, 264, 269, 270
Biota 273
Bloom 273
Brackish 273

C

Carnivore 273
Carotenoid 273
Cell Cycle 99–104
  cell division 99–100
    asexually based division 99
    cell division in eukaryotes 100
    cell division in prokaryotes 100
      archaea 100
      bacteria 100
      binary fission 100
    sexually based division 99
  centrosomes 101
  cytokinesis 101
  how nature controls cell cycles 102–104
    cancer 103–104
      benign tumors 103
      metastasis 103
    external controls 103
    internal controls 103
  importance of cell cycles 102
    G1 phase 102
    typical cell cycle 102–103
      G2 phase 102
      M phase 102
      S phase 102
  mitosis 100–101

anaphase 101
interphase 100
metaphase 100
prometaphase 100
  microtubules 100
prophase 100
telophase 101
nature controls cell cycles
  cancer
    malignant tumors 103
Cell plate 273
Cell, The 3–10
  basic cell structure 7–10
    genetic material 9–10
      chromosome 9
      deoxyribonucleic acid (DNA) 10
      DNA 10
      plasmids 10
    genetic structure
      nucleus 10
    macromolecules 7
    membranes
      cytoplasm 8–9
      hydrophilic lipid 8
      hydrophobic lipid 8
      lipids 7
      plasma membranes 7
      proteins 7
      selective permeability 7
  chemicals to cells 5–6
    atoms and elements 5
    carbohydrates 6
    carbon 5
    fats 6
    hydrogen 5
    molecules and compounds 6
    nitrogen 5
    oxygen 5
    phosphorus 5
    proteins 6
    sulfur 5
  common ancestors 3–4
    bacterial cells 3–4

biological diversity 4
biota 3
blue whale 4
eukaryotes 4
prokaryotes 4
types of cells
archaea 6
eukaryotes 6
membranes 7–8
prokaryotes 6
Chemoreceptor 273
Chitin 273
Chromosome 273
Codon 274
Colony 274
Continental shelf 274
Cytoplasm 274

D

Deciduous tree 274
Deoxyribonucleic Acid (DNA) and Chromosomes 89–97
Cells' DNA 89–92
complementary strands 91
adenine 91
cytosine 91
guanine 91
thymine 91
Crick, Frances 90
deoxyribonucleic acid 89
double helix 90
Franklin, Rosalind 90
genes 91–92
genotype 91
phenotype 91
transposons 92
genetic alphabet 91
genetic code 91
triplets 91
nitrogen base 90
nucleic acid 89
phosphate groups 90
Watson-Crick model 90

Watson, James 90
Wilkins, Maurice 90
chromosomes 94–97
alleles 95
crossing over 96
diploid set 96
genetic variation 97
haploid cells 96
homologous chromosomes 95
meiosis 96–97
sperm and egg 96
spermatozoa 97
traits 95
complementary strands
genetic alphabet
codons 91
Replicating DNA 93–94
antiparallel orientation 94
correct and complementary base parings 94
DNA ligase 94
DNA polymerase 93
helicase 94
primase 94
template 93–94
topoisomerase 94
Domain 274
Dormant 274

E

Ecology 253–260
biodiversity 258–259
climate
optimal range 259
ecological diversity 258
ecosystems 258
extinct species 258
genetic diversity 258
threatened species 258
community 254
deforestation 259
desertification 259–260
energy pyramids 256
habitat fragmentation 259

how species interact 255–256
human activities and ecology 254–255
  climate change 255
  conservation 255
  environmental legislation 255
  global warming 255
  noise and light 255
  population spread and urbanization 255
  renewable energy 255
  resource recycling 255
  waste and pollution 255
  waste reduction 255
human role in Earth ecology 253–254
  biological communities 254
  habitat fragmentation 254
  imbalance 254
  industrialized society 254
nutrients in Earth's ecosystems 256–257
  carbon cycle 257
  iron cycle 257
  microbes 256–257
  nitrogen cycle 257
  nitrogen fixation 257
  oxygen cycle 257
  phosphorus cycle 257
  photosynthesizing (carbon dioxide) 257
  respiring (oxygen) 257
  sulfur cycle 257
population 254
prey and predators 257
species 254
Electron 274
Electron transport chain 274
Environments 245–251
  Earth's environments 246–250
   air 246
   soil 246–248
    deep sediments 247
    deep soils 247
    surface soil (topsoil) 247
   soil environments 248
    leaf litter 248
    parent material 248

    subsoil 248
    topsoil 248
   water 246–247
    deep freshwaters 247
    deep marine waters 247
    surface freshwaters 247
    surface marine waters 247
   water environments (ocean) 248
    abyssal zone 249
    bathyal or mesopelagic zone 249
    benthic zone 248
    coastal zone 249
    continental shelf 249
    euphotic zone 249
    hydrothermal vents 249–250
    limnetic zone 248
    littoral zone 248
    open sea zone 249
    profundal zone 248
   water environments (rivers) 249
    floodplain zone 249
    source zone 249
    transition zone 249
  ecosystems 250–251
   nutrient flow 250–251
    food chain 251
    food web 251
  habitat 246
  hydrothermal vents 250
  microenvironment 247
  niche 246
Enzyme 274
Eukaryote 274
Eukaryotic Cell, The 27–35
  flagella 29
  organelles 28
   centrosome 29
   chloroplast 29
   endoplasmic reticulum (ER) 29
   Golgi apparatus 29
   lysosome 29
   mitochondrion 29
   nucleus 28

peroxisome 29
ribosome 28
vacuole 29
protists 30–34
algae 31–33
amoeba 30
brown
seaweed and kelp 32
Ciliates
Paramecium 31
diatoms 33
dinoflagellates
red tide 30–31
diplomonads
Giardia 30
Euglena 30
first green plants 34–35
golden
carotenoids 32
green
Ulva or sea lettuce 32
Volvox 32
oomycetes
water molds 31
phytoplankton 35
Plasmodium
malaria 31
protozoa 34–35
red
phycoerythrin 32
singled-celled eukaryotes 30
Trichomonas vaginalis 30
water molds, white rusts, and downy mildews
33–34
the first eukaryotes 28
DNA 28
peptides 28
phospolipids 28
prokaryotic cells 28
non-photosynthetic types 28
photosynthetic types 28
RNA 28
Evolution and Extinction 129–135

evolution and biodiversity 134
behavioral adaptation 134
physiological adaptation 134
structural adaptation 134
extinction 134–135
endangered species 135
speciation 135
types of extinction 135
background extinction 135
mass depletion 135
mass extinction 135
Wilson, Edward O. 135
the origin of species 132–133
theory of evolution 129–132
carbon dating 133
catastrophism 130–131
Cuvier, Georges 130
Darwin, Charles 130
HMS Beagle 130
Darwin's theory of evolution 132
de Lamarck, Jean-Baptiste 131
use and disuse 131
descent with modification 132
evolution 132
fossils 132
gradualism 131
Hutton, James 131
Linnaeus, Carolus 133
Lyell, Charles 130–131
macroevolution 133
The Origin of Species 130
biodiversity 130
Extremophile 274

F

Fermentation 71–76, 274
energy without oxygen 72
acetone-butanol fermentation
Clostridium bacteria 74
aerobic organisms 72
alcohol 72
alcohol fermentation
Saccharomyces bacteria 74

winemaking and brewing 75
Zymomonas bacteria 74
anaerobic microbes 72
anaerobic organisms 72
anaerobic prokaryotes 72
butanediol fermentation
   2,3-butanediol 74
   ethanol 74
   formic acid 74
butyric acid fermentation
   dietary fiber 74
carbon dioxide (CO2) 73
cellulose 72
dihydroxyacetone fermentation
   Gluconobacter 74
fermentative organism 72
glycolysis 73
Homoacetate Fermentation
   acetic acid 74
lactic acid 72
lactic acid fermentation 75
   heterolactic fermentation 75
   homolactic fermentation 75
   lactate 75
   Lactobacillus 74
   Streptococcus 74
malolactic fermentation
   Lactobacillus 74
methane (CH4) 73
methanogens 73
mixed acid fermentation 76
   acetic acid 76
   digestive tract 74
   propionic acid 76
   rumen 76
   ruminant 76
propionic acid fermentation
   fiber digestion 74
protozoa 72
pyruvates 72
sorbose fermentation
   Acetobacter 74
   ascorbic acid (vitamin C) 74

Gluconobacter 74
terminal electron acceptor 72
yeasts 72
Fertilization 274
First Cells on Earth 11–18
   amino acids and proteins 16–17
   alanine 16
   amino group 16
   arginine 16
   asparagine 16
   aspartic acid 16
   carboxyl group 16
   cysteine 16
   essential amino acids 17
   glutamic acid 16
   glutamine 16
   glycine 16
   histidine 16
   hydrogen 16
   isoleucine 16
   leucine 16
   lysine 16
   methionine 16
   phenylalanine 16
   proline 16
   protein structure 17
      primary 17
      quaternary 17
      secondary 17
      tertiary 17
   serine 16
   threonine 16
   tryptophan 16
   tyrosine 16
   valine 16
  biochemistry 12
  carbon 12
   carbon dioxide 13
      oxygen 13
      photosynthesis 13
   enzymes 13
   fats 13
  first organic molcules

carboxyl 14
fatty acids 14
first organic molecules 14–15
amino acids 14
glycerol 14
nucleic acids 15
nucleotides 15
phosphate 14
sugars 14
glycogen 13
hydrocarbons 13
coal and oil 13
fossil fuels 13
lipids 13
polymers 13
proteins 13
chemical reactions 11
deoxyribonucleic acid (DNA) 18
first cell 18
fatty compounds 18
nucleotides 18
peptides 18
RNA 18
self-replication 18
how chemicals lead to life 15–16
amino group 16
carboxyl group 15
functional groups 15
hydroxyl group 15
phosphate group 15
sulfhydryl group 16
water solubility 16
Miller, Stanley 12
nitrogen 15
ribonucleic acid (RNA) 17–18
base 18
nucleic acid 18
phosphate 18
ribose 18
Urey, Harold 12
Urey-Miller experiments 12
Flowers, Fruits, and Pollination 225–231
angiosperm life cycle 230–231

anthers 230
bloom 231
endosperm 230
female gametophyte (egg) 230
fertilization 230
haploid microspore 230
male gametophyte (sperm cell) 230
meiosis 230
ovary 230
seed coat 230
seedling 230
stigma 230
tube cell 230
zygote 230
classifying flowers 228
complete flowers
petals 228
pistil 228
sepals 228
stamens 228
imperfect flowers
male or female reproductive organs 228
incomplete flowers 228
perfect flowers
female reproductive organs 228
male reproductive organs 228
fruits 229–230
pericarp 229
rind 229
seed dispersal 230
zygote 229
insects and pollination 229
bumble bees 229
butterflies 229
flies 229
honey bees 229
miner bees 229
nectary 229
pollen grains 229
wasps 229
yellow jackets 229
Parts of a Flower 225–227
angiosperm 225

carpels
  ovary 226
  pistil 226
  stigma 226
  style 226
flower seasons 227
nectary
  nectar 227
petals 225
receptacle 227
sepals 225
  calyx 225
stamens
  anther 227
  pollen 226–227
stigma
  pollen 226
pollination 228
  fertilization 228
  self-pollination 228
Food chain 274
Functional group 274

G

Gene expression 274
Genetic code 274
Genetic Information from Genes to Proteins
      105–110
  gene expression 105–106
    ribosome 107
    RNA and transcription 106
    RNA's job in translation 107
    RNA transcription
      messenger RNA (mRNA) 106
      ribosomal RNA (rRNA) 106
      transfer RNA (tRNA) 106
    RNA translation
      mRNA 107
  genetic code 107–108
    codons 107–108
  ocgx
    elongation 108
    nucleotides 108

polypeptide 109
promoter 108
RNA polymerase 108
transcription unit 109
other components of gene expression 108–109
protein 109–110
  amino acids 109
  denaturation 109
Genetics and Inheritance 111–119
  chromosome gene maps 118
  genes linked on chromosomes 117–119
    Drosophilia 117
    fruit flies 117
    sex-linked genes 118–119
      color blindness 119
      Duchenne muscular dystrophy 119
      hemophilia 119
      meiosis 118
      sex chromosomes 119
  heredity, genetics, and inheritance 112
    evolution 112
    genotype 112
  law of independent assortment 114–115
  Mendelian genetics 112–117
    diploid cell 113
    following traits through generations 115–116
      pea plants 115–116
    gamete 113
    genetic testing 116
    haploid cell 113
    law of independent assortment
      meiosis 115
    law of segregation 114
      allele 114
      paired (homologous) chromosome 114
    meiosis 113
    Mendel, Gregor 112–114
    mitosis 113
    pea plants 112–117
    recessive and dominant traits 116–117
Global warming 274
Globulin 274
Glycogen 274

Glycolysis 275
Gradient 275

H

Herbivore 275
High-energy phosphate bond 275
Homologous chromosome 275
How Cells Communicate 47–53
    cell signaling 51–52
        cytoplasmic determinants 51–52
        induction 52
    chemotaxis 48–49
        attractants 48
        chemoreceptors 48–49
        feedback 48–49
        repellants 48
    controlling gene expression 53
        differentiation 53
        DNA 53
        embryo 53
        proteins 53
        transcription 53
        translation 53
    multicellular organisms 50–51
        bacterial colony 50–51
        cell differentiation 51
        fruiting bodies 50
        myxobacteria 50–51
    neurons 52–53
        neurotransmitters 52
    quorum sensing 49
        chemoreceptors 49
        prokaryotes 49
How We Organize Species 137–143
    DNA homology 141
        annealing 141
    hierarchies 138–141
        domains 139–140
            archaea 139
            bacteria 139
            eukarya 139
        kingdoms 140
            animalia 140

fungi 140
monera 140
plantae 140
protista 140
        taxonomy 140–141
    humans and taxonomy 141–143
        Class Mammalia 142
        Domain Eukarya 141
        Family Hominidae 142
        Genus Homo 142
        Infraclass Eutheria 142
        Kingdom Animalia 141
        Order Primates 142
        Phylum Chordata 142
        Species sapiens 142
        Subclass Theria 142
        Suborder Anthropoidea 142
        Subphylum Vertebrata 142
        Superfamily Hominoidea 142
    phylogeny, systematics, and taxonomy 137–138
        Linnaeus, Carolus 138–139
        species 138
How Your Cells Work 37–45
    biological membranes 38–42
        crossing the membrane 40
            active transport 40
            diffusion 40
            endocytosis 40
            facilitated diffusion 40
        endoplasmic reticulum 38
        energy generation in membranes 40–42
            aerobic organisms 41
            electron transport chain (ETC) 41
            membrane potential 41
            osmosis 41
        how membranes help cells communicate 42
            cell sorting 42
            identification of foreign cells 42
        ion balance 38
        lipid bilayer 39–40
            lymph 39
            permeability 39
        microvilli 39

mitochondria
  chromosomes 43
  mitochondrial membrane 38
  nuclear membrane 38
  plasma membrane 38
cytoskeleton 44
Endoplasmic Reticulum (ER) 43–44
  rough ER 44
  smooth ER 44
  transport vesicles 44
Golgi apparatus 44
lysosomes 45
mitochondria 42–43
  adenosine triphosphate (ATP) 42–43
  cristae 43
  DNA 43
  electron transport chain 43
  mitochondrial matrix 43
nucleus 43
peroxisomes 44–45
Hydrothermal vent 275

I

Imbalance 275
Infectious disease 275
Inside a Cell 8, 9, 15, 20, 22, 30, 31, 34, 39, 41, 43,
        44, 49, 52, 67, 82, 93, 95, 106, 107, 119, 141,
        151, 162, 168, 169, 186, 216, 217, 237, 238
Integumentary system 275
Invertebrates 175–181
  characteristics of early organisms 176–177
    sponges 176
  exoskeleton 179
  invertebrates: animal-like 178–181
    Acanthocephala 179
    Annelida 179
    Arthropoda 179
    Brachiopoda 178
    Chordata 180–181
      hagfishes 180
      lancelets 180
      tunicates 180
    Ctenophora 179

Cycliophora 179
Echinodermata 180
Ectoprocta 178
Hemichordata 180
Mollusca 179
Nematoda 179
Nemertea 178
Onychophora 179
Phoronida 178
Platyhelminthes 178
Priapula 179
Tardigrada 179
invertebrates: nonanimal-like 177–178
  Cnidaria 178
  Kinorhyncha 178
  Loricifera 178
  Placazoa 177
    Trichoplax adhearans 177
  Porifera 178
  Rotifera 178
symmetry 177

L

Lipid 275
Lymph 275

M

Membrane 275
Metabolic pathway 275
Metabolism 57–64, 275
  anabolism 57
  anabolism and catabolism 61–62
    amino acids 61
    carbohydrates 61
    chemical energy 61
    equilibrium 62
    fats 61
    free energy 62
    glycogen 61
    glycolysis 62
    kinetic energy 61
    Krebs cycle 62

metabolism 61
potential energy 61
protein 61
sugars 61
thermal energy 61
triglycerides 61
ATP 62–64
  adenine 63
  adenosine diphosphate (ADP) 63
  ATP-ADP conversion 64
  ATPase 63
  ATP regeneration 63
  ATP synthase 63
  high-energy phosphate bonds 63
  hydrolysis 63
  kilocalories (kcal) 63
  nitrogen base 63
  phosphate groups 63
  respiration 63
  ribose 63
catabolism 57
enzymes 57–61
  amino acids 58
  amylase 58
  coenzymes 59–60
  cofactor 59
  DNAase 58
  how enzymes run chemical reactions 58
    laws of thermodynamics 58–59
  lipase 58
  maltase 58
  metabolic pathway 58
  phosphorylase 58
  protease 58
  regulating enzymes 59–61
    allosteric regulation 60
    chemical work 60
    competitive inhibition 60
    enzyme's active site 59
    feedback inhibition 60
    mechanical work 60
    noncompetitive inhibition 60
    pH 61

temperature 60
    transport work 60
  sucrase 58
Methane 275
Methanogen 275
Microenvironment 275
Microtubules 275
Mutation and Other Genetic Errors 123–127
  mutation 123–124
  mutation leads to adaptation 125
    adaptation 125–126
    genetic variation 126
  mutation rate 125
  mutations
    codons 123
  natural selection 127
    Darwin, Charles 127
    gene pool 127
  types of mutations 124–125
    causes of mutations 124
    deletion 124
    frameshift 124
    insertion 124
    substitution 124

N

Neutron 276
Nitrogen fixation 276
Nucleic acid 276
Nucleus 276

O

Optimal range 276

P

Parasite 276
Pathogen 276
Peptide 276
Peptidoglycan 276
Permeable 276
Photosynthesis 77–84
  chloroplasts 77, 80–82

columns grana 81
light and photosynthetic pigments 81–82
  bacteriochlorophylls a and b 82
  carotenoids 81
  chlorophyll a 81
  chlorophyll b 81
  stroma 81
  thylakoids 81
  thylakoid sacs 81
photosynthesis in bacteria 82
  thylakoid membranes 82
photosynthesis' two stages 82–84
  chemiosmosis 82
  dark reactions 84
    Bassham, James 84
    Benson, Andrew 84
    Calvin cycle 84
    Calvin, Melvin 84
    carbon fixation 84
    stomata 84
  fluorescence 83
  light reactions 83
    cytochrome 83
    glucose synthesis 83
    proton gradient 83
photosynthetic organisms 78–80
  algae 78
  autotrophs 78
  cyanobacteria 78–80
    filamentous cyanobacteria 80
  Euglena 78
  green bacteria 78, 80
  green plants and trees 78–79
  heterotrophs 78
  phytoplankton 79
  protists 78
  purple bacteria 78, 80
Phylum (plural: phyla) 276
Plant Body and Plant Growth 211–218
  anatomy of a plant 215–218
    leaves 217–218
      carbon dioxide uptake 217
      cellulose 217

chloroplasts 217
energy storage 217
gas exchange 217
glucose manufacture 217
petiole 218
photosynthesis 217
polysaccharide 217
transpiration 217
water transport 217
plant cell walls 216
roots 215
  rhizosphere 215
  root hairs 215
stems 216–217
  bulbs 216
  cellulose 216
  lignin 216
  phloem 216
  polysaccharides 216
  proteins 216
  rhizomes 216
  stolons 217
  tubers 217
  xylem 216
wood 218
early plant life 212–214
  bryophytes (nonvascular) 212–213
    diploid stage 213
    haploid stage 213
    hornworts 212
    liverworts 212
    mosses 212
  gametophytes 213–214
  mosses 213
    peat moss 213
  sporophytes 213–214
  vascular plants
    angiosperms 214
    gymnosperms 214
plant diversity 211–212
  aquatic versus terrestrial 212
  climate and geographic habitat 212
  evolutionary history 212

flowering versus non-flowering 212
    seed-producing versus non-seed-producing
        212
    soft plants versus woody plants 212
    vascular versus nonvascular 212
Plant Sensory and Defense Systems 233–240
    apoptosis 234–235
    circadian rhythms 237
    plant defenses 238–240
        herbivores 239
        immune hormones
            salicylic acid 240
        pathogens 239–240
            antimicrobial compounds 240
            natural resistance 240
            phytoalexins 240
            salicylic acid 239–240
    plant hormones 236–238
        abscisic acid 237
        auxin 236
        brassinosteroids 237
        cytokinins 237
        ethylene 237
        fruit growth and ripening 237–238
            ethylene 238
        gibberellins 237
    seed germination 238
    stimuli sensed by plants 234–236
        drought and flooding 234
        gravity 234
        light intensity and direction 234
        mechanical forces 234
        plant's signal response system 235–236
            de-etiolation 236
            photoreception 235
            phytochromes 235
            reception 235
            receptor 235
            response 235
            transcription and translation 235
            transduction 235
        relative humidity 234
        salt levels in soil and water 234

        temperature 234
Plasmid 276
Polymer 276
Polypeptide 276
Polysaccharide 276
Primary producer 276
Prokaryote 276
Prokaryotic Cell 19–26
    bacteria and archaea 19–20
        aerobe 20
        aerobic environments 20
        anaerobic environments 20
        bacilli 20
        cocci 20
        extremophile 21
        methane gas 20
        peptidoglycan 20
        pseudomurein 20
        spirila 20
        spirochetes 20
        vibrios 20
    diversity in prokaryotes 23–25
        cyanobacteria 25
            blue-green algae 25
            large aggregates 25
            long filaments 25
            photosynthesis 25
            single cells 25
        extremophiles 24–25
            acidophiles 25
            alkaliphiles 25
            archaea 24
            barophiles 25
            extreme hyperthermophiles 25
            halophiles 25
            hydrothermal vents 24
            hyperthermophiles 24
            psychrophiles 25
            radio-tolerant 25
            thermophiles 24
            xerophiles 25
        pathogens 23
    domains of living things 26

domain archaea 26
domain bacteria 26
domain eukarya 26
  algae 26
  amphibians 26
  animals 26
  fungi 26
  insects 26
  invertebrates 26
  plants and trees 26
  protozoa 26
  reptiles 26
inside prokaryotes 21
  carboxysomes 21
  cytoplasm 21
  gas vesicles 21
  magnetosomes 21
  photosynthetic vesicles 21
  ribosomes 21
  storage granules 21
outside prokaryotes 21–23
  capsules 22
  Diplobacilli 23
  Diplococci 23
  fimbriae 22
  flagella 22
  Gram, Christian 21
  Gram stain 21
  infectious disease 21
  mycolic acid 22
  pili 22
  polysaccharide layers 22
  Sarcinae 23
  S-Layers 22
  slime layers 22
  Staphylococcus 22–23
  Streptobacilli 23
  Streptococci 23
  Tetrad 23
Protein 276
Proton 276

R

Reaction 276
Red tide 276
Resin 276
Respiration 65–70, 276
  ATP output 70
    ETC 70
    glycolysis 70
    Krebs cycle 70
  energy from organic fuel 65–69
    aerobic organism 66
    anaerobic organism 66
    ATP synthase 69
    Calvin cycle 68
    carbon compounds 65
    carbon dioxide 68
    chemiosmosis 69
    cytochrome 68
    electron transport chain (ETC) 68–69
      chemiosmosis 69
      cytochromes 68
      Mitchell, Peter 69
      mitochondria 68
      proton 69
    glucose 66
    glycolysis 66
    heme 68
    hemoglobin 68
    Krebs cycle 67
      acetyl coA 67
      acetyl coenzyme A 67
      citric acid cycle 67
      glycolysis 67
      Krebs, Hans Adolf 67
      tricarboxylic acid cycle 67
    phosphorylation 69
      oxidative phosphorylation 69
    redox 66
    reduction and oxidation 66
  oxygen
    glycolysis 70

hemoglobin 70
  respiratory pigments 70
  terminal electron acceptor. 70
respiratory syncytial virus
  viral diseases
    nervous system 170
      AIDS (HIV) 170
      rabies (rabies virus) 170
      viral encephalitis (encephalitis virus) 170
    urinary-genital tract 170
      genital warts (human papillomavirus) 170
Ribosome 277
Ruminant 277

S

Seeds and Plant Life Cycles 219–224
  diploid form 220
  gametophytes 220
  haploid form 220
  life cycle of seed plants 223–224
    diploid zygote 223
    fertilization 223
    germination 223
    megasporangium 223
    meiosis 223
    ovulate cones 223
    photomorphogenesis 223
    pollen cones 223
    pollination 223
  seeds 221–222
    atropine 222
    integument 221
    megasporangium 221
    megaspore 221
    menthol 222
    micropyle 221
    morphine 222
    pollen 221
    taxol 222
  sporophytes 220
Single-Celled Organisms: Bacteria and Archaea
    147–154

archaea 148–150
  archaea without cell walls 149–150
    Ferroplasma 150
    Thermoplasma 150
  halophiles 149
    Halobacterium 149
    Halococcus 149
  methane producers 148–149
    Methanobacterium 149
    Methanococcus 149
    methanogen 148–149
    Methanosarcina 149
  sulfur users 148
bacteria 150
  flagella 150
  green nonsulfur bacteria 153
  green sulfur bacteria 153
  mycobacteria 153
    Mycobacterium tuberculosis 153
  mycoplasma
    ureaplasma 154
  peptidoglycan 151
  photosynthetic bacteria 152
    Chlorobi 152
    Chloroflexi 152
    Cyanobacteria 152
  proteobacteria 152
    Beggiatoa 152
    Campylobacter 152
    E. coli 152
    Rhizobium 152
    Salmonella 152
    Shigella 152
    Vibrio 152
  purple nonsulfur bacteria 153
  purple sulfur bacteria 153
  spirochetes 153
    Borrelia 153
    Leptospira 153
    Treponema 153
  thermoacidophiles 150
Sterol 277

symbiosis 277
Symbiotic relationship 277

T

Template 277
Today's Biology 267–272
  biology specialties 268
    anatomy 268
    biochemistry 268
    cell biology 268
    developmental biology 268
    evolution 268
    genetic engineering 268
    genetics 268
    genomics 268–270
      genome sequencing 270
    immunology 268
    microbiology 268
    molecular biology 268–269
      bioengineering 269
    parasitology 268
    physiology 268
    proteomics 268, 270–272
    systems biology 268
  conservation biology 272
  gene therapy 271
Trait 277
Transmission 277
Tundra 277

V

Viruses 165–172
  bacteriophages
    phages 166
  capsid 166
  Ebola virus 167
  how viruses infect 168–170
    lysogenic cycle 169–170
    lytic cycle 169
  human immunodeficiency virus (HIV) 169
  influenza 168
  nonliving particles in biology 170–171

prions 170–171
  bovine spongiform encephalopathy 171
  chronic wasting disease 171
  Creutzfeldt-Jakob disease 171
  fatal familial insomnia 171
  feline spongiform encephalopathy 171
  Gerstmann-Sträussler-Scheinker syndrome
    171
  kuru 171
  mad cow disease 171
  scrapie 171
  viroid 170
outer coating 166
polio 167
retrovirus 169
  acquired immunodeficiency syndrome (AIDS)
    169
Tobacco Mosaic Virus (TMV)
  Beijerinck, Martinus 167
  Ivanowsky, Dimitri 167
  Mayer, Adolf 167
  Stanley, Wendell 167
viral diseases 170
  lower respiratory tract
    viral pneumonia (respiratory syncytial virus)
      170
  upper respiratory tract
    cold (Rhinovirus) 170
    flu (influenza) 170
viral genetic material 166
virus classification 167
  capsid shape and symmetry 168
  disease caused 168
  DNA or RNA size 168
  infected organism 168
  lipid envelope 168
  method of replicating 167
  transmission method 168
  type of genetic material 167
virus size
  submicroscopic 167